Roger McNeill

The Prevention of Epidemics

And the Construction and Management of Isolation Hospitals

Roger McNeill

The Prevention of Epidemics
And the Construction and Management of Isolation Hospitals

ISBN/EAN: 9783337022105

Printed in Europe, USA, Canada, Australia, Japan

Cover: Foto ©berggeist007 / pixelio.de

More available books at **www.hansebooks.com**

THE
PREVENTION OF EPIDEMICS

AND THE

CONSTRUCTION AND MANAGEMENT
OF ISOLATION HOSPITALS

BY

ROGER MᶜNEILL, M.D. Edin.

D.P.H. Camb.

MEDICAL OFFICER OF HEALTH FOR THE COUNTY OF ARGYLL, FORMERLY RESIDENT MEDICAL OFFICER
INFECTIOUS HOSPITALS OF THE METROPOLITAN ASYLUMS BOARD AT HOMERTON AND
DEPTFORD, LONDON, AND H.M.S. 'ATLAS' AND 'ENDYMION,' GREENWICH

WITH ILLUSTRATIONS

LONDON

J. & A. CHURCHILL,

11 NEW BURLINGTON STREET

1894

PREFACE

THE nature of infectious diseases, the manner in which they spread, the loss and misery caused by them, and the measures that should be adopted to prevent them, are but imperfectly known to those entrusted with the administration of the laws affecting the public health. Even members of the medical profession who have not made a special study of the question hardly realise its importance.

In the following pages an attempt is made to deal with the subject in so far as it affects small towns and rural districts. In the principal towns every endeavour is now being made to cope with this difficult problem, but in many small towns and rural districts little in the way of preventive measures has hitherto been attempted. And it has, perhaps, not yet been sufficiently realised that as long as infectious diseases are allowed to spread in the small towns and rural districts, so long will it be found to be impossible to prevent epidemics from breaking out in more populous centres.

Local authorities have now for many years been in existence, whose duty it is to devise means against the spread of disease. Yet in many districts little has been done beyond appointing medical officers of health to advise Local Boards, and sanitary inspectors to carry out their instructions. The necessary appliances have not, as yet, been provided. It would be as rational to attempt to put down crime by merely appointing experts who should investigate the career of criminals, and write interesting and exciting narratives to the magistrates and judges regarding them, or by a force of police without the lock-up and the dungeon behind them, as it is to appoint medical officers of health to investigate the origin and spread of infectious disease, and sanitary inspectors to act on their instructions, without providing proper appliances for the isolation and disinfection of houses and patients.

The designation " Isolation Hospitals" has been adopted on the title-page in preference to either " Fever" or " Infectious Hospitals." A " Fever Hospital" does not convey a pleasant idea to the

public mind. Too often have such buildings been
erected during a period of panic, and in the face of
a threatening danger. Isolation Hospitals are still
regarded by the public as a source of danger to
persons living in their vicinity. It is not known that
such buildings may be constructed and managed in
such a way as to be of no danger whatsoever. The
existence of an Isolation Hospital in a locality has not
yet come to be regarded by the public as a guarantee
that proper measures are being taken against the
spread of infection, or the absence of such a building
as a sign that persons suffering from epidemic
diseases are allowed to infect the community. A
person suffering from an infectious disease and isolated
in an Isolation Hospital is placed under the most
favourable conditions for his own recovery, and at
the same time, by his isolation, he is prevented from
becoming a source of danger to others. The separation of the infectious sick from the healthy is the
primary object of such hospitals, and the designation
" Isolation Hospital" conveys this idea better than
either " Fever" or " Infectious Hospital." With a
view to prevent any feeling among the well-to-do
against being removed to a hospital, it might be

advisable to adopt such a designation as "Sanatorium" or "Isolation Home." This has already been done in some parts of England.

My thanks are due to Dr. Thorne Thorne, Medical Officer to the Local Government Board, for assisting me to get information regarding several hospitals, and for his kind permission to copy hospital plans from his report ; to the architects and other gentlemen mentioned in the appendix, for supplying me with plans and photographs of hospitals, as well as other information regarding them ; and to Dr. M'Naughton, Medical Officer for the county of Kincardine, for assisting me in revising the proof sheets.

ROGER M'NEILL.

OBAN, 23rd Oct. 1894.

CONTENTS

CHAPTER I

THE DISSEMINATION OF INFECTIOUS DISEASES

CHAPTER II

THE INFLUENCE OF EFFECTIVE MEASURES AGAINST THE SPREAD OF INFECTION

CHAPTER III

THE GAIN TO THE COMMUNITY THROUGH THE ADOPTION OF EFFECTIVE MEASURES AGAINST THE SPREAD OF INFECTION

CHAPTER IV

The Nature of Infection and its Influence on the Construction and Management of Isolation Hospitals

CHAPTER V

The Establishment and Erection of Isolation Hospitals—Preliminary

CHAPTER VI

Hospital Construction

CHAPTER VII

Disinfection

CHAPTER VIII

Remarks on Hospital Management

CHAPTER IX

PRIVATE SANITARY AID ASSOCIATIONS

Evidence by Mrs. Francis Johnstone before Smallpox and Fever Hospitals Commission—Rules of the Sanitary Aid Association at Hastings Pages 204-213

CHAPTER I

THE question of devising proper means for the isolation and treatment of cases of infectious diseases is at present prominently before the public mind, and in Scotland is under the direct consideration of almost every District Committee and County Council. It is of even more than national importance. Hitherto the subject has nowhere received the attention it deserved. Under the old form of Local Government in Britain it was scarcely possible to make much advance in this direction. Until a comparatively recent period the nature of infectious diseases and the manner in which they spread were but little inquired into. It was believed that "the pestilence that walketh in darkness and the destruction that wasteth at noon-day" originated and spread far beyond human ken. That "prevention is better than cure" was a byword, but its scope was not fully understood. Persons with weak chests knew that certain precautions prevented them from catching cold; the rheumatic were aware that living in a damp climate or in damp houses should

B

be avoided; and persons of a bilious temperament
knew that adherence to a certain diet and regimen
prevented the frequency of their attacks, but the fact
that various diseases might be prevented if not extinguished on a large scale, was not fully grasped.
Gradually it began to be known that medical men who
devoted their time and energy towards the improvement of the public health were doing something to
prevent the population from dying at the same rate as
was formerly the case.   Newspapers drew attention to
the fact that in the large towns where more attention
was paid to this question, the rate of mortality was
gradually getting lower.   It is not yet, however,
sufficiently well known that the lowering of the rate of
mortality in the towns is due to a large extent to the
measures taken to prevent the more fatal of the infectious diseases from spreading.   The improvement
in the sanitary condition of the houses of the poor, the
prevention of overcrowding, the care taken to provide
pure water and unadulterated food, and the advance
in the general wellbeing of the population, have
directly improved the public health, and have, furthermore, rendered persons less liable to suffer from, and
better able to withstand attacks of, infectious and other
diseases.   Yet, if measures were not taken to prevent
the spread of infection in the towns, the mortality
would be much greater than is now the case.   In
rural districts, until a comparatively recent period, this
question was almost entirely neglected, and even yet,
over a great part of the country, measures are carried

out in a very imperfect and slipshod fashion. Even to-day the advent of an epidemic sometimes results in a panic. Unseemly huts, meant for Isolation Hospitals, defective in structure and general arrangements, are sometimes hurriedly run up during the progress of an epidemic. Upon the advent of an epidemic a proper site for a hospital can rarely be obtained. Buildings hurriedly erected are often found in very unsuitable localities, and frequently present a forbidding appearance. They have been badly built, warmed, and ventilated, far from any other dwelling, and sometimes even without a road. Hitherto it was in very rare instances indeed that a proper ambulance carriage was provided for the conveyance of patients to a hospital. An old rickety jolting cab, in which a patient could only sit, was considered quite suitable, although it rendered the conveyance of severe cases from a distance extremely dangerous, if not impossible. It cannot therefore be wondered at, that persons whose friends or relatives suffered from infectious diseases hesitated to send them by such means into such buildings. As might be expected, medical men hesitated to advise their patients to take advantage of such provision.

The inefficiency of the provision for the isolation of infectious diseases in rural districts was, to a certain extent, due to want of knowledge on the part of Local Authorities and partly to the smallness of the ratable area under each Parochial or Local Board. A little parish could hardly afford to provide a separate Isola-

tion Hospital, while local jealousies and want of cohesion
frequently prevented the erection of joint or combination hospitals.   This state of matters has been considerably improved, more particularly in the case of Scotland,
by the creation of larger districts, with District Com-
mittees as Local Authorities, more or less under the
eye of the County Councils.   In 1879 the Local Government Board got a return of Isolation Hospitals in
England.[1]   Of 1593 Sanitary Authorities in England
and Wales it was found that only 296 had provided
means for the isolation of infectious diseases.   In many
instances the hospital provision reported was such in
name only.   A large number of Isolation Hospitals
have been erected in England since that time.   Still,
"notwithstanding the phenomenal growth of public
opinion in regard to this question in recent years, many
important centres of population in the provinces and as
yet a large proportion of the smaller towns, as well as
the majority of rural communities, have not moved
in the matter."[2]   In a paper read in the Architectural
Section of the Seventh International Congress of
Hygiene and Demography, held in London in 1891,
Dr. Thorne stated : "I find that since that date (1881)
loans amounting in all to £448,769 have been granted
to over a hundred Sanitary Authorities for the provision
of Isolation Hospitals . . . and I am probably correct
in stating that some adequate means for the isolation

[1] 10th Report of Local Government Board, 1880-81. Supplement to p. 5,
reissued 1893.
[2] Burdett, *Hospitals and Asylums of the World*, vol. iii. p. 102.

of infectious diseases are now possessed by at least 400 urban, rural, and port Sanitary Authorities."

Under the Public Health Act,[1] Local Authorities may provide within their districts hospitals for the reception of the sick for the use of the inhabitants. Provision is also made whereby Local Authorities are authorised to "borrow for the purpose of so building or otherwise providing permanent hospitals on the security of the general assessments."[2] Money for this purpose can be borrowed at a low rate of interest payable by small instalments extending over a number of years.

Wherever a large population is crowded together, as in the principal towns, where the individual members are liable to come into daily contact, and where frequent communication exists with other populous places, infectious disease of one kind or another almost always prevails. The seeds of infection in such places are probably never extinct. Clothes, articles of furniture, and parts of premises in which infectious disease prevailed, sometimes escape thorough disinfection and cleansing. Persons who are unfortunate enough to come in contact with these articles, or to visit or live in such houses, are liable to become infected. These infect others, and so disease spreads. The movements of the individual members of the community are so devious and intricate, the communica-

---

[1] See Sect. 131, P. H. Act for England, 1872 ; and Sect. 39, P. H. Act for Scotland, 1867.

[2] P. H. Scotland Amendment Act, 1871 ; English P. H. Act, 1875.

tion between different persons so frequent and intimate,
the seeds of infection so tenacious of life, and so liable
to stick to all manner of things, that unless the utmost
care is taken infectious disease will spread from a
small centre and do infinite mischief.[1]   The misery
following epidemics of infectious diseases among the
poorer classes of the population cannot well be calcu-
lated.   The rate of mortality from infectious disease is
not equally distributed among all classes of the population.   The poor, living in unhealthy dwellings, and badly
nourished, suffer most severely.   In many cases where
a wage-earning member of a family dies, the survivors
are brought to the verge of starvation, and thus become
dependent, perhaps for many years, for subsistence on
parochial relief.

Only those who make a special study of this question can be fully impressed with its importance.   In
towns, people know only their immediate neighbours.
In rural districts the population is often sparse, and
the means of communication slow.   The sufferings of
a family in one locality are heeded but for a brief
space of time, and by their immediate neighbours

---

[1] The following cases, recorded by Dr. Birdwood, medical superintendent to
the hospital ships, Long Reach, Kent, illustrate this.   A dock labourer was the
first of a group of ten patients admitted into the smallpox hospital ships, London.
He did not know the source of his infection.   His brother, a little girl, and a
man living in the same house, were infected.   A man not employed by any
sanitary authority came to disinfect the house, and a woman from Camberwell
came to visit her dying daughter at the same house ; they were both admitted
suffering from smallpox.   The potman at the public house frequented by the
[dock labourer] was the earliest one to recognise the nature of the disorder with
which he was afflicted.   The potman, his wife, a fellow-lodger, and a lad who
daily called for beer, followed.   See Report of Statistical Committee of Metropolitan Asylums Board, 1891, p. 10.   Numerous other instances might be given.

only. Infectious diseases have visited them periodically for ages, and much as they suffer and dread these, they look upon them as part of the natural order of things. It seldom occurs to any one that such **misery** and death might be entirely prevented.

The prevalence of infectious disease **in the large** towns is a grave danger to all places in frequent **communication** with them.[1] Isolated **cases keep breaking**

[1] In 1892 Dr. M'Lintock, county medical officer for Lanarkshire, reported a fatal case of smallpox at Newmains, and stated that the source of infection was ultimately, and with difficulty, traced by Mr. Dobson (the sanitary inspector) to a daughter living in Glasgow, who had been visiting her parents a fortnight previous to the rash on her mother (Second Annual Report, p. 65).

Dr. Maxwell Ross, county medical officer, Dumfriesshire, states that the parishes of Annan, Kirkpatrick-Fleming, and Middlebie, suffered most from [scarlet fever] in the Annan district, and that some of the cases were introduced from Hoddam. The Annan hiring fair appeared to be responsible for certain of the cases which occurred both in Annan and Kirkpatrick-Fleming. Into the same district measles was introduced from Carlisle (First Annual Report, pp. 17 and 20).

Dr. M'Vail, county medical officer for Dumbartonshire, states, in regard to an outbreak of scarlet fever on board the training ship *Empress* in the Gareloch : "The first case was that of a new boy who had come from Glasgow on September 10th;" and in regard to diphtheria in Cumbernauld parish, "the disease appeared to have been brought from Edinburgh" (First Annual Report, pp. 51 and 54).

Dr. Bruce, county medical officer for Ross-shire, states, in regard to an outbreak of measles in the camp of the Inverness-shire Militia, at Muir of Ord : "The origin of the epidemic was distinctly traced to Harris, and to the West it unfortunately was carried afresh" (First Annual Report, p. 40).

Dr. Ogilvie Grant, county medical officer for Inverness-shire, referring to outbreaks of a "very severe and malignant type of measles" in Harris, North Uist, and Skye, states : "Infection was traced by Dr. Mackenzie to two sources —(1) The disbanding of the Inverness-shire Militia ; (2) Direct infection from the adjoining parish of Harris. . . . Dr. Dewar of Portree states that one of the crew of the s.s. *Lochiel* was infected from another source, viz. from a militiaman returning home from Muir of Ord, who was travelling in the eruptive stage of the disease" (First Annual Report, p. 11).

Dr. Watt, county medical officer for Aberdeenshire, states, with regard to scarlet fever at Newhills : "Many of these cases were traced to infection spreading from the town, where, from August to December, a very severe epidemic was raging" (First Annual Report, p. 8).

In 1892 I reported that a person suffering from "typhoid **fever at** Innellan **is** believed to have become infected while examining old **properties**

out every year in localities visited by excursionists from the towns at the various holiday seasons.

Unless every care is taken, each individual case is liable to start an epidemic.

The towns, however, are not alone at fault. The rural districts and smaller burghs also infect the larger · towns. Epidemics of typhoid fever [1] have frequently been traced to infected milk supplied from the country. Epidemics of scarlet fever have been traced to the same source. [2] The emigration of persons from infected houses in rural districts during the progress of epidemics also frequently infects the towns. It will thus be seen that the towns and the rural districts, in

in Glasgow, where the disease is said to have prevailed." . . . " Measles was introduced from Newcastle to Kilmun, and from Rothesay to Ormidale " . . . " And from Glasgow to Inverary " . . . " An outbreak of typhoid fever in the island of Jura was traced to a young man who came home after suffering from the disease in Glasgow " (Second Annual Report, pp. 50 and 56, and First Annual Report, p. 7).

[1] An epidemic of typhoid fever in Plymouth was traced by Dr. F. M. Williams, medical officer of health, to the milk supplied from a farm in Plympton district (*British Medical Journal*, 1892, p. 1157). Typhoid fever was traced to consumption of lemonade, ices, and spirits, which have been made of, or diluted with, water from a sewage-polluted well (Mr. G. H. Fosbroke, *Treatise on Hygiene and Public Health*, vol. ii. p. 324: Stevenson and Murphy). An epidemic of typhoid fever at Whitchurch in 1871 was traced to the introduction of the poison by a young person who came into the place ailing, and ultimately died of typhoid fever (Report of Medical Officer Local Government Board, 1872).

[2] An outbreak of scarlet fever, associated with diphtheria and sore throat, was traced by Dr. Parsons to the milk supply in Macclesfield rural and urban sanitary districts (Report Medical Officer Local Government Board, 1889, p. 84). Scarlet fever in Marylebone, London, in 1885-86, was traced by Mr. Power to milk supplied from Hendon (Report by Dr. Klein in the Report by the Medical Officer of the Local Government Board, 1887, p. 203). An epidemic of scarlet fever in Glasgow in August 1892 was traced by Dr. Russell to milk supplied from a dairy farm in Renfrewshire (Report on an Outbreak of Scarlet Fever in Glasgow, by Drs. J. B. Russell, A. K. Chalmers, and E. Klein, 1893). An outbreak of diphtheria at Yorktown, Camberley, was traced by Mr. W. H. Power to the milk supply (Report of Medical Officer Local Government Board, 1886, p. 311). Numerous other instances might be given.

so far as the spreading of infectious disease is concerned, act and react on each other.

Innumerable centres scattered over a wide area in rural districts and villages converge upon the large towns and infect them. On the other hand, from the towns, thousands of infected cases radiate over the country, infecting individuals and creating new centres from which the disease may again spread and return to them. In the towns " couriers arrive bestrapped and bebooted, bearing joy and sorrow in pouches of leather ; there, top-laden, with four swift horses, rolls in the country baron and his household ; here, on timber legs, the lamed soldier hops painfully along begging alms. A thousand carriages and wains and cars come tumbling in with food, with young rusticity, and other raw produce, animate or inanimate, and go tumbling out again with produce manufactured " ; and thus infectious diseases as well as other commodities are exchanged.

Dr. Russell,[1] Medical Officer of Health for the City of Glasgow, states :—" In my experience epidemics never cover the whole area of the community at their incidence. They begin in one district, and if not checked, they eat their way through the mass, while, if they are vigorously attacked, they may be stamped out in that district. If they begin within the municipal limits [of Glasgow], the whole repressive force of the municipality is directed to the spot. If they begin

[1] *Transactions of the Epidemiological Society of London*, 1881-82, pp. 84 and 85.

outside these limits, the appliances of the petty burgh, even at their best, cannot command adequate resources, and we can only stand by to quench such sparks as may be projected into our premises."

The towns, however—more particularly the large towns—have as a rule done their part in trying to provide ways and means to prevent the spread of infection within the town and beyond its borders. From the very nature of the case, the lead in preventing the spread of infectious disease would begin there. There the evil was ever present, and the amount of wealth centred within a limited area justified the expenditure of large means to cope with it. The question was not so pressing in rural districts. Infectious disease here spreads more slowly, but its victims will be picked out all the same. It will thus be seen that, however well the towns are managed and guarded in this respect, they cannot be kept free of infection so long as they are supplied with disease from the country. Remedial measures for country districts and small towns, therefore, become essential, not merely for the protection of the rural population, but of the urban population as well.

Infectious diseases may be compared to a fire, and the population to the combustible material. In towns the combustible material is lying in heaps, and when the spark falls the spread and destruction are more rapid.

But even in more sparsely-peopled districts the material is sufficiently continuous to enable the flame to spread, if with slower, yet with equally fatal steps. In the case of infectious disease, the spark cannot be immediately extinguished, as may be done in the case of a fire ; but fortunately it can be removed or isolated. The patient may be separated from the healthy if a case occurs in a house of sufficient accommodation, and if not he can be removed to a hospital, if such accommodation is provided.

CHAPTER II

THE INFLUENCE OF EFFECTIVE MEASURES AGAINST THE SPREAD
OF INFECTION ON THE EXTENT AND FATALITY OF FUTURE
EPIDEMICS

MANY of the infectious diseases prevalent among the
community are diseases of children, and therefore many
of the adult population are protected by a previous
attack.   It might accordingly be argued that if children
are prevented from getting such diseases, the population will after a time become more liable to suffer from
more severe epidemics, that a larger number will be
liable to catch those diseases, and that it will become
impossible to prevent them from spreading.   No doubt
this is true to a certain extent, but meantime our
organisation for the prevention of disease will become
more perfect.   The centres from which such diseases
spring will become more isolated and less numerous.
The sanitary condition of the houses of the poor will
be better.   Scientific men are not only engaged in
tracing the source of such diseases and in framing
measures to prevent them from spreading, but great
strides have already been made in devising means
whereby the population may be rendered almost proof
against even the most infectious of them.   Vaccination

and re-vaccination have almost extinguished smallpox.
Dr. Watt found that the deaths from smallpox were
reduced to one-fifth of their original number by vaccination. Before 1800, 20 out of every 100 born,
excluding still-born, perished by this dreadful malady.[1]
This disease is no longer a periodical scourge decim-
ating the population. Even if it breaks out in isolated
localities, it can easily be kept under control. This has
been amply proved in Scotland during the past year.
Smallpox broke out in various places in the country,
but in no instance was it allowed to spread into anything like an epidemic. Typhus fever is no longer so
prevalent as it once was. It only occurs in isolated
spots where the rules of health are neglected, where
human beings are badly fed, in overcrowded, dirty
dwellings without sufficient ventilation. Pasteur has
been successful, by means of inoculation, in protecting
many from an attack of hydrophobia. By the same
means animals have been rendered immune to the
action of the anthrax bacillus. Infectious diseases are
thus as it were between two fires. In front we have
our local authorities, with their army of medical officers
of health and sanitary inspectors doing their utmost to
check the onward march of these diseases, while at
the rear we have bacteriologists attacking them in
their very citadel, not without hope of being able
ultimately to extinguish them one by one. In addition
to this, by the gradual improvement in the wellbeing

---

[1] Farr, *Vital Stat.*, p. 322. *Enquiry into the Relative Mortality of the Principal Diseases of Children in Glasgow*, by Dr. Watt, 1813.

of the population, by the providing of purer water, healthier food, and more sanitary dwellings, epidemics of some diseases appear to be less fatal or of a milder type than was formerly the case.

It might also be argued, that infectious diseases are milder in children, and that if children are prevented from contracting these diseases, as adults they may get them in severer forms and die in proportionally higher numbers. This is true to a certain extent, and in a very marked degree, of typhus fever. If one hundred children under five[1] years of age suffered from this disease, about seven of them would die ; if the same number aged from ten to fifteen were to get it, only between two and three would die ; if the hundred were about thirty years of age, about from twenty to thirty would prove fatal ; if about fifty, fifty-four would die ; and if over sixty, sixty-seven would succumb ; yet no one has yet had the courage to recommend that efforts should be made to spread typhus fever among children. It is no longer universally prevalent, and the localities from which it emanates are so few that it is possible to check its spread. As age advances the liability to catch typhus fever decreases, although the disease is more fatal in adults.

The succeeding table[2] gives the admissions of persons suffering from typhus fever, the percentage of admissions, and rate of mortality, at different ages

[1] Murchison, *Continued Fevers*, p. 236. Report of Statistical Committee of Metropolitan Asylums Board, 1890.

[2] Report of Statistical Committee of Metropolitan Asylums Board, 1890, p. 14.

in the Hospital of the Metropolitan Asylums Board
from 1872-1890.

TYPHUS FEVER

Hospital of the Metropolitan Asylums Board, 1872–1890

Ages.	Admissions.	Percentage of Total Admissions.	Deaths.	Mortality per cent of Admissions at Same Age.
Under 5	86	4.05	2	2.32
5–10	242	11.40	1	0.41
10–15	370	17.44	14	3.80
15–20	352	16.59	28	8.04
20–25	239	11.26	49	20.50
25–30	152	7.16	34	22.36
30–35	159	7.49	47	29.74
35–40	129	6.08	47	36.43
40–45	168	7.92	81	48.21
45–50	95	4.47	42	44.21
50–55	60	2.82	36	60.00
55–60	32	1.50	24	75.00
And upwards }	37	1.74	27	72.97
Totals .	2121		432	20.36

The following table[1] gives the same data in regard
to the admissions into the London Fever Hospital
from 1848-1870.

[1] Murchison, *Continued Fevers,* 3rd ed., p. 237.

London Fever Hospital, 1848–1870

Ages.	Admissions.	Deaths.	Percentage of Total Admissions.[1]	Mortality per cent of Admissions at Same Age.
Under 5	234	15	1.29	6.69
5–9	1196	43	6.59	3.59
10–14	2189	50	12.06	2.28
15–19	2932	131	16.16	4.46
20–24	2400	248	13.23	10.33
25–29	1727	262	9.52	15.17
30–34	1518	312	8.36	20.55
35–39	1458	378	8.03	25.92
40–44	1507	464	8.30	30.79
45–49	1039	442	5.72	42.54
50–54	790	392	4.35	49.62
55–59	441	238	2.42	53.96
60–64	400	241	2.20	60.25
65–69	188	142	1.03	75.53
70–74	84	61	.46	72.62
75–79	32	27	.17	84.37
80 and upwards	3	3	.01	100.
Age not specified	130	8		6.15
Totals .	18,268	3457		18.92

In typhoid fever, again, the death-rate among children under ten years of age is less than half that among adults over thirty. Children are, however, very liable to contract this disease if they come in the way of infection, whereas persons of maturer years are less susceptible to it. That is to say, if an equal number of children and of persons of mature years were exposed equally to the infection of typhoid fever, the vast majority of the children would contract the disease, whereas the vast majority of the adults would suffer no

[1] In reckoning the percentage of admissions at different ages, the 130 cases in which the age is not specified are deducted from the total.

harm.   Of 6960 cases of typhoid fever admitted into the Metropolitan Asylums District Board Hospitals before 1890, 2718 were under fifteen years of age, whereas only 255 were over forty.

As age advances the liability to catch typhoid fever appears to decrease.   " In children the liability to invasion from greater susceptibility rather than from greater exposure to it causes a maximum proportion of attacks." [1]

The succeeding table [2] gives the admissions of persons suffering from typhoid fever, the percentage of admissions, and rate of mortality, at different ages in the hospitals of the Metropolitan Asylums Board from 1872-1890.

Ages.	Admissions.	Percentage of Total Admissions.	Deaths.	Mortality per cent of Admissions at Same Age.
Under 5	201	2.88	28	13.93
5–10	894	12.84	77	8.61
10–15	1623	23.31	217	13.37
15–20	1604	23.04	295	18.38
20–25	1054	15.14	211	20.02
25–30	689	9.89	163	23.65
30–35	401	5.76	107	26.68
35–40	239	3.43	63	26.36
40–45	126	1.81	28	22.23
45–50	72	1.03	22	30.56
50–55	33	.47	11	33.34
55–60	12	.17	6	50.00
And upwards }	12	.17	4	30.34
Totals .	6960		1232	17.70

[1] Farr's *Statistics*, p. 391.
[2] Report of Statistical Committee of Metropolitan Asylums Board, 1890, p. 13.

C

The following table gives the same data in regard to the admission into the London Fever Hospital[1] from 1848 to 1870 :—

Ages.	Admissions.	Percentage of Total Admissions.[2]	Deaths	Mortality per cent of Admissions at Same Age.
Under 5	58	.98	7	12.06
5– 9	558	9.44	63	11.28
10–14	1174	18.16	151	12.86
15–19	1588	26.86	246	15.48
20–24	1164	19.69	238	20.36
25–29	600	10.15	123	20.50
30–34	297	5.36	76	25.59
35–39	201	3.40	53	26.36
40–44	124	2.09	33	26.61
45–49	64	1.08	14	21.87
50–54	36	.60	8	22.22
55–59	20	.33	9	45.00
60–64	20	.33	9	45.00
65–69	5	.08	2	40.00
70–74				
75–79	2	.03	1	50.00
Age not specified	77	1.30	1	1.29
Totals .	5988		1034	17.26

Nearly half the cases were between 15 and 25 years of age, more than ¼th under 15, ¼th above 30, while only 1 in every 71 exceeded 50.

Of 7348 cases of typhoid fever reported to the French Academy from different parts of France, 2282, or 31 per cent, were under 15 years of age.[3]

If now, scarlet fever be considered, it will be found that the disease gets less fatal to a very marked degree as age advances.   From an analysis of 42,111

[1] Murchison's *Continued Fevers*, 3rd ed., pp. 437 and 438.
[2] In reckoning the percentage of admissions at different ages, the 77 cases in which the age is not specified are deducted from the total.
[3] Gaultier de Claubry, 1849, xiv. 29.

cases admitted into the Metropolitan Asylums Board
Hospitals, it will be seen that while children under five
died at the rate of twenty in every hundred; the death-
rate of the patients over fifteen years of age was only
about 4 per cent. The Statistical Committee aptly
remarks :[1] "Such results are sufficient to prove how
essential it is that every precaution should be taken to
prevent the exposure of young children to infection,
and they effectually dispose of the once popular notion
among ignorant people that it is best to suffer from the
disease while young."

The following table gives the admissions of persons
suffering from scarlet fever at different ages, and the
rate of mortality of 42,111 cases admitted into the
hospitals of the Metropolitan Asylums Board from
1871 to 1891 :[2]—

Ages.	Cases Admitted.	Percentage of Total Admissions.	Deaths.	Mortality per cent of Admissions at same Age.
Under 5	12,077	28.67	2407	19.93
5–10	17,423	41.37	1175	6.74
10–15	6,994	16.60	238	3.40
15–20	2,859	6.78	95	3.33
20–25	1,428	3.39	49	3.43
25–30	661	1.56	27	4.08
30–35	371	.88	21	5.66
35–40	168	.39	11	6.55
40–45	71	.16	6	8.45
45–50	31	.07	1	3.23
50–55	21	.04	1	4.76
55–60	2	.004	1	50.00
And upwards }	5	.01	1	20.00
Totals .	42111		4033	9.58

[1] Report of Statistical Committee, 1890, p. 11.　　[2] Ibid. 1891, p. 24.

With regard to diphtheria it will be seen that this disease is also more fatal in children than adults. The Statistical Committee of the Metropolitan Asylums Board states, after analysing 3075 cases treated in their hospitals, that "diphtheria, like scarlet fever, is most fatal to infant children. The maximum mortality rate is obtained in the second year of life, when it reaches the high percentage of 68.79, subsequently falling with every additional year of life to the minimum of 1.05 per cent amongst persons between twenty-five and thirty years of age." Not only is the disease more fatal in children than adults, but it would also appear that more children, in proportion, suffer from the disease, whether from greater susceptibility or from a more continuous exposure to the influences under which the disease arises. It cannot be argued that adults are protected by previous attacks of this disease, for one attack of it does not protect a person from future attacks.

The following table [1] gives the number of admissions of persons suffering from diphtheria at different ages, the percentage of the total admissions, the deaths, and the rate of mortality in the hospitals of the Metropolitan Asylums Board from 1888 to 1891, and for N.W. Hospital for 1892 :—

[1] Report of Statistical Committee of Metropolitan Asylums Board, 1891. Figures for N.W. District Hospital supplied by Dr. Gayton, Med. Supt.

NUMBER of ADMISSIONS of Persons suffering from Diphtheria at different ages, the percentage of the total admissions, the deaths, and the rate of mortality, in the hospitals of the Metropolitan Asylums Board from 1888 to 1891, and for N.W. Hospital for 1892.

Ages.	Admissions M. A. Bd. 1888-91.	N.W. Dist. Hospital, Admissions 1892.	Total.	Percentage of Total Admissions.	Deaths.	Mortality per cent of Admissions at Same Age.
Under 1	57	18	75	2.00	41	54.66
1–2	173	28	201	5.38	130	64.67
2–3	257	47	304	8.14	185	60.85
3–4	335	60	395	10.58	197	49.87
4–5	334	67	401	10.74	176	43.89
Total under 5	1156	220	1376	36.86	729	52.97
5–10	951	206	1157	30.99	351	30.33
10–15	378	63	441	11.81	51	11.56
15–20	210	70	280	7.50	12	4.28
20–25	163	43	206	5.51	12	5.82
25–30	95	24	119	3.18	3	2.52
30–35	52	16	68	1.82	3	4.41
35–40	30	4	34	.91	2	5.88
40–45	20	8	28	.75	2	7.14
45–50	11	2	13	.34	3	23.07
50–55	3	0	3	.08	1	33.33
55–60	5	1	6	.16	1	16.66
And upwards	1	1	2	.05	2	100.00
Totals .	3075	658	3733		1172	31.39

The following table[1] gives the average annual number of deaths from diphtheria at certain ages to every 100,000 living at those ages in England and Wales, 1861 to 1870 and 1871 to 1880:—

[1] See *Diphtheria*, by Dr. Thorne Thorne, p. 38.

Average annual number of Deaths from Diphtheria at certain ages to 100,000 living at those ages in England and Wales, 1861 to 1870, and 1871 to 1880.

Decennia.	All Ages	0-1	1-2	2-3	3-4	4-5	5-10	10-15	15-20	20-25	25-35	35-45	45-55	55-65	65-75	75-
1861–70	18	58	91	80	83	73	39	14	6	4	3	2	2	3	3	2
1871–80	12	29	49	48	58	55	29	9	3	2	2	2	1	2	2	1
Mean	15	43	70	64	70	64	34	11	4	3	2	2	1	2	2	1

It will thus be seen that more children suffer from diphtheria than adults, while at the same time it is, like scarlet fever, most fatal in its effects in children under five years of age.

For measles and whooping - cough statistics are meagre. Although these diseases cause a greater number of deaths than any of the other infectious diseases, no efficient provision has yet been made to prevent them from spreading. Dr. George Blundell Longstaff, in his *Studies in Statistics*, states : " Measles causes nearly five times, and whooping-cough more than six times, as many deaths as smallpox, while the amount of mischief to be attributed to these little-thought-of ailments in the way of impaired general health, permanent lung disease, and even blindness and deafness, will probably never be known. It would appear that preventive medicine has failed to control these diseases. Can this be because the people will not co-operate to put down diseases which affect only young children ?"

The following statistics show that, in the case of measles, the gross mortality is highest in infancy and childhood. Of 9532 deaths from measles in London from 1861 to 1872, 8566 were under four years of age, and of 458 cases treated in the London Fever Hospital during the years 1878 to 1882, there were seventeen deaths, thirteen of which were of children under five years of age.[1] These figures do not give the number of cases and deaths at different ages, and therefore no

[1] Collie on *Fevers*, pp. 211 and 214.

conclusion can be arrived at in regard to the severity of this disease at different periods of life. Nor are there any reliable statistics to show the susceptibility to catch the disease at different ages. Few authorities have apparently paid attention to the subject. Dr. Thomas,[1] however, states that "measles are essentially dangerous to young children. Its danger decreases rapidly with accession of years, and in the late years of manhood is already at a minimum. In old people, who have, however, but little predisposition and are rarely attacked, the disease is again dangerous." According to Passow[2] the absolute mortality from measles in Berlin from 1863 to 1867 increased up to the second year of life, at which point it reached it greatest height. From the third year on it diminished, at first rapidly and then slowly, up to the thirtieth year; not constantly, however, since in the eight and tenth year there was a slight increase, while from the twentieth to the twenty-fifth year no deaths took place. From the thirtieth to the thirty-fifth year the mortality again increased slightly ; above thirty-five there died only one person (aged sixty-two). In an epidemic at the Faroe Islands in 1846, alluded to by various authorities, where the disease had not prevailed since 1781, and where, therefore, probably the whole population under sixty-five years of age were liable to suffer, the gross rate of mortality for the first nine months of the year was nearly three times more than the average in the first

[1] Zeimssen, *Cyclopædia of the Practice of Medicine*, vol. ii. p. 112.
[2] *Ibid.* p. 112.

year of life. From one to twenty the mortality was
about the normal, from twenty to thirty it was 1.4
times more, from thirty to forty, 2.4 times; forty to fifty,
2.6 times; and from fifty to sixty, 4.5 times. Passow
reports that in this epidemic "not one old person
previously unaffected exposed to infection escaped,
although with some younger individuals this was the
case." Copland[1] states that the susceptibility probably decreases with the progress of age. In regard
to the malignant type of measles Aitken[2] is of opinion
that the danger is greater in the old than in young
people.

The following table, compiled from data given in
Zeimssen's *Cyclopædia of Medicine*, vol. ii. pp. 112-113,
gives the number of cases of measles treated at Vienna
in 1864-67, at Meerane in 1861, and at Frankfort
in 1860-61, and the deaths at various periods of life.

---

[1] *Practice of Medicine*, p. 813.   [2] *Ibid.* vol. i. p. 460.

Table, compiled from data given in *Zeimssen's Cyclopædia of Medicine*, vol. ii., pp. 112-113, giving the number of cases of Measles treated at Vienna in 1864-67, at Meerane, in 1861, at Frankfort in 1860-61, and the deaths at various periods of life.

Ages.	Cases.				Deaths.				Percentage of deaths.
	Vienna, 1864-67.	Meerane, 1861.	Frankfort, 1860-61.	Total Cases.	Vienna.	Meerane.	Frankfort.	Total.	
Under 1	16	112	45	173	6	2	8	16	9.2
1— 2	35	221	156	412	21	19	15	55	13.3
2— 3	52	264	204	520	26	26	9	61	11.7
3— 4	47	236	186	459	13	7	3	23	5.0
4— 5	39	204	243	486	10	6	4	20	4.1
5—10	183	482	954	1619	22	3	7	32	1.9
10—15	...	227	...	227	...	1	...	1	0.4
15—20	...	...	...	...	...	...	...	...	...
Total.	372	1736	1788	3896	98	64	46	208	5.33

In the case of whooping cough statistics are still more meagre. West[1] states that age exerts an evident influence on the mortality of whooping cough as well as on its prevalence, both being greatest in early childhood. Bristowe,[2] quoting Dr. Smith, states that whooping cough is the most fatal of all children's diseases under one year of age, 68 per cent. of all the deaths from it occurring under two years. According to Newsholme,[3] the mortality from whooping cough is highest in the first year of life. It then diminishes like that of measles, but more rapidly, becoming quite insignificant by the 10th year. Steffen[4] gives West's tables, and states that the younger the children the more dangerous the disease. Fagge[5] is of opinion that when this disease affects adults it is very distressing, but is not dangerous nor of long duration. I know myself of an island with a population of 381 at the last census, where no case of whooping cough occurred for twenty years until 1892. On visiting the island in October of that year, I found 114 persons suffering from the disease, at all ages up to 20. The population being scattered, some families entirely escaped. There was no death. Here only a few cases were very young children. On the other hand, Dr. Russell of Glasgow informs me, that of 266 cases of whooping cough treated at the Belvidere Hospital in 1892, 23.5 per cent. died, but he adds that this

---

[1] *Diseases of Infancy and Childhood*, p. 475.
[2] *Theory and Practice of Medicine*, p. 148.
[3] Newsholme's *Statistics*, p. 182.   [4] Zeimssen, *Cyclopædia*, p. 717.
[5] *Principles and Practice of Medicine*, p. 1138.

cannot be taken as the general mortality, "as our patients were mostly poor, ill-nourished, often otherwise diseased children."

Excepting the fact that measles, like typhus fever, is dangerous in the case of the aged, it may be taken as proved, that both measles and whooping cough diminish in severity from childhood to the verge of old age. If, therefore, the public would see the propriety, indeed the duty, of preventing the "slaughter of the innocents" that is at present going on, it might be left to future generations to adopt such precautions as would protect the aged. The love of self-preservation would then make the aged raise their voice, whereas at present the victims are dumb, or their cry gets no hearing.

In smallpox, the rate of mortality, as well as the susceptibility to the disease, is affected rather by vaccination than by age. Of 10,403 cases treated by Dr. Gayton at the Homerton Hospital, 2085 had good vaccination marks. The mortality was 3 per cent, 4854 had imperfect vaccination marks, and these died at the rate of 9 per cent. In 1295 cases they were stated to be vaccinated, but no vaccination mark was visible; 27 per cent of these died. Of 2169 unvaccinated cases 43 per cent proved fatal. The following table gives the number treated, the deaths, and the rate of mortality, at different ages in the four classes :—

Table giving the number of cases of Smallpox at different ages treated by Dr. Gayton, in the hospitals of the Metropolitan Asylums District Board, distinguishing those with good vaccination marks, imperfect marks, vaccinated but without visible marks, and those not vaccinated (10,403 cases).[1]

Ages	Vaccinated with good marks.			Vaccinated with imperfect marks.			Vaccinated left with no visible marks.			Not vaccinated.		
Years.	Cases.	Deaths.	Per cent.	Cases.	Deaths.	Per cent.	Cases.	Deaths.	Per cent.	Cases.	Deaths.	Per cent.
0–2	4	0	0	32	3	9	22	9	41	276	181	66
2–5	57	0	0	150	18	12	96	38	40	401	202	50
5–10	206	2	1	532	27	5	207	40	19	510	180	35
10–15	439	5	1	939	32	3	214	42	20	317	74	23
15–20	606	12	2	1037	66	6	205	39	19	204	86	42
20–25	389	11	3	843	109	13	167	56	34	174	83	48
25–30	189	12	6	529	80	15	116	35	30	105	56	53
30–40	147	14	10	526	78	15	137	49	36	103	42	41
40–50	29	4	14	186	33	18	85	24	28	49	21	43
50–	19	2	11	80	18	22	46	20	43	30	13	43
All ages	2085	62	3	4854	455	9	1295	352	27	2169	938	43

[1] *Vaccination Vindicated*, by Dr. Gayton, p. 88.

It will be observed that in those " vaccinated with good marks," the number of very young children affected was small and the mortality nominal, whereas in the unvaccinated, under five years of age, the mortality and susceptibility were excessive. On the other hand, while the susceptibility shows considerable diminution between the ages of ten and fifteen, the mortality during this period was not only much less than among the infants, but even less than at any subsequent period of life.

It will thus be seen that in the case of all the infectious diseases prevalent[1] in this country, adults are less liable to catch them than children, and, with the single exception of typhus fever, that these diseases are most fatal during infancy and childhood. The prevention of infectious diseases would therefore not cause the population to suffer from more extensive or more severe epidemics (although at rarer intervals) in the future. On the contrary, the extension of isolated outbreaks into epidemics would be prevented with the same ease as isolated outbreaks of typhus fever are prevented from spreading into general epidemics at the present day. Not so long ago, typhus fever prevailed over the whole country, and consequently the resources existing were not able to cope with it. Now, if an outbreak occurs in any locality, it is immediately suppressed. This disease now breaks out only at comparatively rare intervals in isolated localities, and

---

[1] Smallpox, because of the protective influence of vaccination, can hardly be said to be prevalent in this country now.

local authorities with the means at their disposal are able to check its spread. If the other infectious diseases were, by means of Isolation Hospitals and other measures, once got under control, and the population protected in the early years of life, a larger number would, no doubt, be liable to suffer, as they would not be protected by a previous attack, exactly in the same way as the whole population is, at the present day, liable to suffer from typhus fever. This, however, would be more than compensated for by the fewer outbreaks that would occur, and by the greater resources and increased facilities for dealing effectually with them. Although a higher proportion of the population would be liable to suffer, outbreaks would be rarer and could be more easily kept under control. Even if such outbreaks did occur, the diseases would not have the same tendency to spread with rapidity among persons at all ages as they have among children, and fewer would die. I have already shown that the susceptibility to suffer from these diseases gets less as age advances, and, with the exception of typhus and typhoid fevers, the rate of mortality also decreases.

Even with the imperfect means hitherto provided for preventing the spread of infection, considerable progress has been made. In the case of typhus fever, there has been an almost continuous decrease in the number of deaths in England and Wales for the past twenty-four years. The number of deaths from that disease fell from 4281 in 1869 to 318 in 1885. During the same period, the number of deaths from typhoid

and simple continued fevers decreased from 13,967
in 1869, to an average of 8657 per annum, during the
five years 1876-80, and 6671 per annum from 1881 to
1885, and this notwithstanding the increase of the
population.   The deaths from scarlet fever, at all ages
during 1851-60, amounted to 88 per 100,000 of the
population in England and Wales.   From 1861-70 the
rate rose to 97.   From 1871-80 only 72 per 100,000
died.   Since 1878 the rate fell continuously until 1886,
when it was only 17 per 100,000.   From 1838-42
the number of deaths from smallpox was 57.2 per
100,000, from 1865-69 it fell to 14.4.   From 1870-74
it rose to 42.7 per 100,000, but fell from 1875-79 to 8·3,
and from 1880-84 to 6·5 per 100,000.[1]   If energetic
measures were carried out, and efficient means provided by local authorities for the purpose, this progress
would proceed at an accelerated rate until the infectious
diseases most prevalent amongst us would be as rarely
met with as spotted typhus, or the black plague, which
at one time decimated the population.

[1] *Progress of Preventive Medicine during the Victorian Era*, by Dr. Thorne
Thorne.

CHAPTER III

In Scotland the population in 1881 was 3,735,573, and in 1891, 4,025,647. As may be seen from the table on the following page, the number of deaths from the most common of the zymotic diseases during the 10 years was 77,780, giving an average of 7778 per annum.

The economical loss to the community by this item alone is very large. Dr. Farr has shown that each individual member of the community has an actual money value represented by the wages he is capable of earning, and that this constitutes the most important factor in the wealth of the community. Deducting the amount required for subsistence, £150 is the mean nett value of each member of the male population, estimated by the standard of the agricultural labourer. This amount is obtained by capitalising the income derived from wages and deducting all the expenses of subsistence. It is evident that the value will vary greatly at different

D

Number of Deaths from Zymotic Diseases in Scotland for the ten years, 1880-1889

	1880.	1881.	1882.	1883.	1884.	1885.	1886.	1887.	1888.	1889.	Deaths.	Probable Number of Cases.
Smallpox . . .	10	19	3	11	14	39	24	17	3	8	148	888
Measles . . .	1427	1012	1289	1629	1440	1426	681	1598	1406	1948	13,856	...
Scarlet fever . .	2165	1573	1583	1336	1266	944	1058	1179	732	701	12,537	125,370
Diphtheria . .	838	816	961	747	830	688	583	805	872	968	8108	24,324
Croup . . .	801	799	906	952	945	757	668	800	742	914	8284	...
Whooping-cough .	2641	1620	2108	2968	2511	2157	1882	3212	1722	2268	23,089	...
Typhus fever .	170	229	180	152	138	111	80	126	162	69	1417	7085
Enteric . . .	1338	1004	1204	998	1050	889	755	835	665	795	9533	57,198
Relapsing ,, . .	4	...	2	1	2	1	2	7	6	1	26	
Simple Continued do.	155	115	90	71	63	58	62	65	58	45	782	
	9549	7187	8326	8865	8259	7070	5795	8644	6368	7717	77,780	

ages, being partly dependent on the expectation of
life, and very low in infancy on account of the large
death-rate under 5 years of age.

On Dr. Farr's basis the child of an agricultural
labourer is worth only £5 at birth, £56 at the age
of five, £117 at the age of ten, £192 at the age of
fifteen, increasing to £246 at the age of twenty-five,
then steadily declining to only £1 at the age of
seventy, while at eighty the cost of future maintenance
is greater than the earnings by £41.[1]

The 7778 deaths per annum in Scotland from
zymotic diseases occurred at various ages. If, however, the half of that number be taken as males and
multiplied by £150 (the mean value of each member
of the male population), the economical loss to the
community will be found to amount to a considerable
sum. "A still greater economical loss to the community is caused by the preventible sickness and
mortality from fifteen to forty-five years of age.
Trusting to Farr's English Life Tables, of 1,000,000
children born, 72,397 die between the ages of fifteen
and forty-five."[2] . . . " Premature death is an evil and
a loss to the state, but sickness is, from an economical
aspect, a still greater evil ; for sickness adds an

---

[1] Newsholme's *Vital Statistics*, p. 14.

[2] By them [infectious diseases] and by other causes, out of 1000 children
born in Liverpool, 518 were destroyed in the first ten years of their life, some by
smallpox, many by measles, scarlet fever, whooping-cough, typhus and enteric
fever. Out of 1000 children born in London, 351 die under ten years of age
by zymotic diseases and other causes. In the healthy districts of England, out
of 1000 children born, 205 die in the first ten years of life.—Farr's *Statistics*,
p. 327.

additional burden, while death can but remove the bread winner." [1]

The 7778 deaths per annum is therefore far from representing all the loss. Every death represents a group of cases which had suffered from the disease, but had recovered completely or partially. The number that recover from the various infectious diseases varies not only according to the disease, but also according to the age and social position of the sufferer, as well as the type of the special disease and other conditions. In the hospitals of the Metropolitan Asylums District Board, during a period of twenty years, 42,111 cases of scarlet fever were admitted ; of these 4033 died, being at the rate of 9.5 per cent. Of 3733 cases of diphtheria admitted into the same hospitals, 1172 died, giving a mortality of 31.3 per cent. Of 6960 cases of typhoid fever, 1232 died, being at the rate of 17.7 per cent, and of 2121 cases of typhus fever, 432 died, giving a mortality of 20.3 per cent. The rate of mortality in 57,037 cases of smallpox during the same period was 17.3 per cent.

If persons suffering from these diseases died at the same rate in Scotland during the past ten years, it will be seen that each of the 7778 deaths from infectious disease represents a group of persons, varying in number according to the disease, who had suffered but survived. Every death from scarlet fever would represent about ten cases, from diphtheria about three cases, from typhoid fever about six cases,

[1] Newsholme's *Vital Statistics*, p. 279.

from typhus about five, and from smallpox about six cases.

The 148 deaths from smallpox during the ten years would represent about 888 cases of that disease, or about eighty-eight cases per annum. The 12,537 deaths from scarlet fever would represent 125,370 cases, or about 12,537 cases per annum. The 8108 deaths from diphtheria would represent 24,324 cases of that disease, or about 2432 cases yearly, and the 9533 deaths from typhoid fever during the ten years would represent 57,198 cases of that disease, or about 5719 per annum.

The loss incurred, even by the simple interference with labour, through this enormous amount of sickness can scarcely be computed, " and the consequences which are left behind (in the survivors) in the loss of health, activity, sight, hearing, and other faculties which render life happy and useful, are lamentable." [1]

After taking the evidence of the highest authorities, the Fever and Smallpox Hospital Commission reported that the period of seclusion in cases of fever lasts on an average eight weeks, and in smallpox six. In an average-sized house and family it has been found to be impracticable to isolate a person suffering from fever or smallpox effectually from the rest of the inmates. The result is that all susceptible persons living in such houses are liable to catch the disease if one of the members contracts infection. On many occasions it is also impossible to prevent the inmates

[1] Report of Fever and Smallpox Hospitals Commission, p. 29.

from mixing with others and carrying infection to
them. It is also found to be a great hardship to many
a business or working man to have to give up work
and lose his wages at the very time his family is ill,
and when he has to incur extra expenses for medical
attendance, medicine, and food. When a case of
infectious disease occurs in a house, all the inmates
are, as a rule, not immediately attacked. Another
catches the disease from the first case within a few
days of the commencement of the outbreak. The
second case, however, will not fall ill at once. The
number of days varies. Perhaps, as a rule, it would be
safe to reckon it from nine to fourteen days. Other
members of the family may escape longer, and some
may not suffer at all. Taking the average length of
time a person should be secluded or isolated from the
public as eight weeks, it will be seen that even with
two cases of infectious disease in a family, it will take
about ten weeks before the house and clothing can be
disinfected and the inmates allowed to mix with safety
with others. If even one member of the family, with
wages of 2s. 6d. per working day, were kept from
work, either through illness or from the risk of spreading infection, the loss amounts to £7 : 10s. Every
family in the community runs more or less the same
risk of being infected by fevers and incurring this loss,
although the full loss would only be felt when an adult
male was disabled or prevented from earning his
wages. This loss falls in full force on the poorest of
the population, the very class least able to stand the

strain.   Supposing now that an hospital were provided,
the first case in the family immediately removed, and
the house furniture and clothing properly disinfected,
the house and family, instead of being a menace to the
neighbourhood for months, would be rendered perfectly
harmless, and the well-being of the family would be
interfered with to the least possible extent.   They
would be able to attend to their work, and the
chance of others catching the disease would be
lessened to the minimum.   Supposing the valuation
of an area provided with a small isolation hospital
is £82,000, and supposing the erection of such a
hospital would cost a sum of £2000, a little cal-
culation will show that that amount might be paid
back in thirty years' instalments, with interest at $3\frac{1}{2}$
per cent, at a cost to the ratepayers of about one
farthing and a half in the pound.   This small rate
would afford means whereby misery, hardship, and
death would be prevented among all classes of the
population, greater security would be given against
the spread of infection, poverty and pauperism would
be lessened, and school attendance improved.   The
population would gain more by the saving in doctors'
fees, in other losses and expenses incidental to sickness and death from infectious diseases, and in the
decrease of poor rates, than they would lose by their
small contribution towards the building, upkeep, and
maintainance of an isolation hospital.   The yearly
charge would be exceedingly small in proportion to the
benefit that would accrue from such an institution.

I do not know of any more profitable investment.
Money invested in an Isolation Hospital and charged
on the rates may be compared to voluntary contri-
butions to a society to provide against sickness or
accidents.   Members of a Local Authority, such as
Burgh Commissioners, County Councillors, or members of Local Boards, may be compared to the directors
of such a society.   They are elected by the public to
take all necessary precautions against disease and ill-
health, and they are invested by statute with authority
to borrow money and levy rates to meet the necessary
expenditure.

Into the commodious, secluded, and well-appointed
mansions of the rich, infectious disease does not so
easily find its way, and even when it does it may be
possible to isolate a case so as to be of very little
danger to the other inmates, and to enable them to
mix with their neighbours and attend to their voca-
tions with little or no risk of spreading the disease.
The following (and it is only one of many cases which
might be quoted) illustrates one of the many ways
by which infectious disease may spread, and shows
how important it is for the well-to-do, even for their
own personal safety, to do their utmost to provide
proper means for the isolation and safe management
of cases of infectious disease.   Deputy Surgeon-
General Bostock, a member of the Metropolitan
Asylums Board, stated before the Fever and Smallpox Hospitals Commission : " A very remarkable case
occurred two or three days ago which will illustrate

what I say. Six children were admitted on Saturday
night to the Hospital at Stockwell [suffering from
smallpox]. They all belonged to one family and
lived in a small apartment at the top of the Hay-
market. Their father is a tailor working for a first-
rate firm in the West End of London ; and about
three weeks ago one of his sons (there were nine of
them altogether) had smallpox, which was treated at
home and was not notified. . . .—The father was a
tailor, you say ? Yes, in the employ of a fashionable
tailor in the West End of London, and in addition to
this, underneath was a laundry employing five women
who came every day, where washing was taken in
from the neighbouring families."

The following is recorded by the Hon. Rollo
Russell :[1]—" A school teacher returned home from
another place with diphtheria. Within the next six
months cases occurred among her family and relations.
It entered the house of the family physician, and
several deaths occurred. He left the town and his
house was vacant for some months. People went to
and from the infected houses to the post-office, which
was also a grocer's shop, and the grocer visited the
houses with the groceries. His family was next
attacked and broken up, and its members scattered.
A new physician came to occupy the vacant house.
Soon after moving his children were attacked. A
lying-in woman whom he attended and her boy of
seven were both attacked ; then a neighbour who

[1] *Epidemics, Plagues, and Fevers*, p. 138.

called on this woman; and so the disease extended
for eighteen months."

Isolation hospitals do not at once ensure the total
absence of infectious disease. If universally adopted,
however, and full advantage were taken of them,
this desirable result would reach towards completion
in the future, and disease from infection would get
rare indeed. Side by side with Isolation Hospitals,
other measures, such as disinfection of clothing and
premises, would have to be practised, and the sources
from which infection springs would be discovered and
extirpated one by one.

Infectious diseases of various kinds prevailed in
this country for generations, probably since man first
inhabited the British Isles. The *materies morbi* or
the "thing that infects," or by means of which disease
is transmitted from one person to another, is very
tenacious of life, and maintains its power of infection
under certain conditions outside the human body for a
long time. It adheres to clothes, it lodges in cracks
and crevices, or on inequalities of house walls, and some
kinds live in insanitary drains and cesspools. Even
if every precaution be taken by means of isolation and
disinfection for preventing such diseases from spreading
from person to person, it is evident that for a long
time occasional outbreaks will occur, if the infecting
material of any particular disease is disturbed where
it now lies dormant, and if it finds its way into the
human system. By isolation and disinfection, new
centres or premises will be prevented from becoming

infected. The sources or infected premises from which epidemics of some diseases now spring will, through the inquiries and investigations of health officers, be detected and disinfected, and outbreaks of such diseases will get gradually rare in proportion. There will therefore ultimately be no pecuniary loss to the community in being rated for hospitals, while the benefit to future generations will be incalculable.

By the aid of Isolation Hospitals, the enforcement of the provisions of the Public Health and the Infectious Disease Notification Acts, with ordinary care and vigilance on the part of the medical officers of health and sanitary inspectors, infectious diseases may in time be almost entirely blotted out of rural districts. In the ordinary discharge of their duties, houses and localities where such diseases formerly existed, and from which they may again spread, will be gradually discovered, and precautions taken to prevent such a catastrophe. I have myself discovered several localities and houses where typhoid fever and diphtheria broke out time after time. If such places are disinfected and existing sanitary defects remedied, outbreaks of infectious disease will not again originate in them. Without hospitals, however, new centres from which such diseases may spread, multiply in spite of every precaution that can be taken.

The following case illustrates this : A girl from one parish was at service in a farmhouse about ten miles from her home. She took ill in the house of her employer, and was sent home to her father's house.

The case was notified to me as typhoid fever on 7th
September 1892. On inquiry I found that typhoid fever
had broken out several times during the past few years
in the same farmhouse in which this case originated.
One death was registered from typhoid fever at a
neighbouring cottar's house in October 1889, where
several persons suffered from the disease. This outbreak was traced to the same farmhouse. In January
1890 some persons in the farmhouse are reported to
have suffered from some disease resembling fever, but
its nature was not known. In the autumn of the same
year five persons in another neighbouring cottar's
family are reported to have suffered from typhoid
fever. A member of this family had also been at
service in the same farmhouse and was sent home
ill with typhoid fever. The sanitary inspector found
the drainage defective, the house infested with rats,
and an outhouse at the back on a higher level and
draining towards the farmhouse.

The excretions in typhoid fever contain the agent
of infection, probably in the most concentrated condition.
If these are thrown out into old stone-built drains, without
the most thorough disinfection the organism of infection
will lodge and thrive in the filth along the course of
the sewer. In the same manner soil saturated with
organic matter about insanitary dwellings may harbour
infection independent of its position. Lodged in some
positions the infection may remain harmless for years.
If, however, by any chance infected particles of filth
get mixed with water, or milk, or food, and thereby

find their way into the system, the disease is again produced, and perhaps in a more virulent form.

It is probable that some of the numerous epidemics of typhoid fever which have been traced to the milk supply in the various large towns arose from similar centres.

The persons that contracted the fever in the above-mentioned farmhouse were treated at their homes in old insanitary dwellings, instead of being isolated in a hospital. These dwellings may in future act as centres for other outbreaks. Isolation Hospitals would, therefore, tend to prevent the multiplication of infected centres, as well as check the immediate spread of such diseases. Of all diseases cholera is probably the most dangerous in this respect. The *comma bacillus* or infecting agent of that disease appears to live and multiply in soil or water. Wherever the disease is likely to break out, Isolation Hospitals should be provided to prevent it from spreading. The immediate isolation of the first case or cases that appear, the thorough disinfection or destruction of excretions, and of all infected articles and premises, are absolutely necessary. If cases are treated at their own homes, the soil in and about the houses is bound to get more or less infected, wind blows about the infected dust, or water gets infected by the soakage and drainage of the infected soil. Every house in which a case is treated becomes an additional source of danger, and a centre from which the disease may spread.

That hospital provision is the most effective means at present known to prevent the spread of infection may be taken as proved beyond doubt. A little consideration will at once show this. A large proportion of the population live in one or two-roomed dwellings, and in such dwellings it may be taken as impossible to isolate the sick from the healthy.

In Aberdeenshire, of 60,551 families, 8250 families live in houses of one apartment, and 19,891 in houses of two rooms.

In Forfarshire, of 64,795 families, 13,820 families live in one-roomed, and 28,936 in two-roomed dwellings.

In Ayrshire, of 46,874 families, 12,143 families live in houses of one room, and 18,427 families in houses of two rooms.

In Lanarkshire, of 231,633 families, 74,389 live in one-roomed, and 98,126 in two-roomed houses. The figures in all these counties include large towns.

In Inverness-shire, which is mainly rural, of 18,856 families, 2818 families live in houses of one room, and 7192 in houses of two rooms.

In Argyllshire, of 17,442 families, 2502 live in one-roomed, and 6125 families lived in two-roomed dwellings.

In the rural parts of Scotland as a whole, at the last census (1891), 46,371 families lived in houses of one apartment, 109,159 families in houses of two, and 53,125 families in houses of three apartments.

The mean number of persons in each family in

Scotland in (1891) was 4.5, and the number of persons to a house in the rural districts was 5.05.

One may take it as a rule that houses of three apartments and under, more particularly in rural districts, are small, badly built, drained, and ventilated.

When infectious disease breaks out in a family living in such a house, it is practically impossible to prevent the rest of the family from getting it, unless the first case is removed to a hospital.

The following cases, recorded by Drs. Barker and Cheyne, in their *Treatise on Fevers*, illustrate what is liable to happen in the absence of an Isolation Hospital :—

" Previous to the opening of the hospital many instances of extreme misery occurred. I would particularise the following :—E. F., a young woman whose husband was obliged, in order to seek employment, to leave her almost destitute with three children in a miserable cabin, was induced one night to give the shelter of her roof to a poor beggar, who it appears had fever. The consequence was that she caught the disease, and from the terror and alarm created in the neighbourhood, was, with her three children, deserted, except that some persons left a little water and milk in the window for the children : one about four, the other three, and the third an infant at the breast. In this way she continued for a week, when a neighbour heard of her distress and sent her a loaf of bread, which was left in the window. Four days after this he grew uneasy about her, and one night he prepared some tea and bread, and taking a

female servant with him, set off to her relief. When
he arrived the following scene presented itself. In
the window lay the loaf, where it had been deposited
four days previously; in one corner of the cabin, on
a little straw, without covering of any kind, lay the
wretched mother actually dying, and her infant dead
by her side, for want of that sustenance which she
had not to give. On the floor lay two children, to
appearance also dying of cold and hunger. At first
they refused to take anything, and he had to force a
little liquid down their throat, and in a short time they
revived, and with the cautious administration of food,
recovered the effect of their suffering. The mother
expired before the visitor quitted the room, who, I
am happy to add, did not suffer for his humanity."[1]

" Mr. and Mrs. L. kept a boarding-school for young
ladies, and a preparatory school for boys, and Mr. L.
attended private pupils. They had nine children, the
elder of whom were capable of assisting in the school.
Having maintained an irreproachable character, they
had every reason to hope that their exertions would
enable them to support their family in comfort. But
their hopes have been blighted. A servant who had
the prevailing fever introduced the infection into the
family in January 18—, and during nine months, notwithstanding all the precautions that were taken, the
fever has occurred at different periods, during which
time the nine children, four of the boarders, two servants, and last of all, Mr. L. himself,—in all sixteen

[1] Barker and Cheyne on *Fevers*, vol. i. pp. 65 and 66.

persons,—have been attacked. . . . The consequence
has been that the pupils have been removed, and the
private tuition has been discontinued, that debts have
been incurred to support life, and all the means of paying them have been cut off, . . . and a length of time
must elapse after his own recovery before he can
resume private tuition lest he should be instrumental
in bringing upon other families that scourge by which
he has so much suffered." [1]

Only two years ago typhus fever broke out in an
isolated township on an island on the West Coast of
Scotland. Like many other places the Local Authority
had not, and still have not, made provision for the
isolation of cases of infectious disease. The fever
spread, and several persons died, among them an old
man who had volunteered to nurse a case of typhus.
By this time the population got alarmed. No one could
be got to coffin the body until an old pedlar volunteered to assist the widow. After this the widow and
her granddaughter were shunned, and no one would
admit the pedlar into his house for fear of infection.
They tried to leave the district by steamer, but
being infected they were refused. The next port
was about thirty miles away, and there they thought
they would not be known. The panic, however,
spread, and the news preceded them. They could
not get a hire, and so they had to walk. Although it
was winter, with showers of snow, they were refused
shelter on the way. When tired, they had to sleep on

[1] Barker and Cheyne on *Fevers*, vol. i. p. 67.

E

the roadside. When within eight miles of the harbour the old man lay down exhausted. The widow and her grandchild trudged along and reached the port. They were allowed to sleep under an old boat until a steamer arrived, and they went to Glasgow, where both suffered from typhus fever.

The pedlar was found dead by the roadside, so that the brave spirit of humanity which the poor old fellow had shown alone amid his neighbours cost him his life.

On the other hand, in his evidence before the Infectious Hospitals Commission, Dr. Dudfield,[1] medical officer of health for Kensington, stated : " If we hear of a case pretty quickly, and it is removed, we reckon to have done with that house ; we hear no more of the disease there."

Dr. Thorne[2] stated : " In the Alcester Rural District in Warwickshire, I found that the early removal of cases of scarlet fever from houses which contained children who were unprotected by having had previous attacks, had prevented any spread of infection. For example, in three instances the first pupil attacked had been removed from school, and in each of these cases no spread took place ; whereas on another occasion, when scarlet fever attacked a pupil at a school, and it was attempted to treat it in isolation in the school building, the disease spread, and seven other attacks followed in the schoolhouse."

[1] Smallpox and Fever Hospitals Commission, Minutes of Evidence, Question 1596.   [2] *Ibid.*, Question 1129.

Dr. Collier,[1] medical officer of health for Fulham, remarked : " No doubt 70 per cent of the cases that occur, and I say this after the most serious reflection, could be prevented, if every case was disclosed. In my reports published last year, I have shown that where cases of smallpox have been immediately reported, and the patients immediately removed, not a single person has caught the disease at the same house, whereas in houses where the patients have not been removed other cases have occurred."

Dr. Browning,[2] medical officer for Rotherhithe, stated, after detailing a case : " Here was a case of a well-to-do tradesman, a boot and shoemaker, a hot-headed anti-vaccinationist, who lost his wife and three children in succession, by the most aggravated form of smallpox, who persisted in refusing to allow the others in the house to be revaccinated, and refused to allow them to be sent to hospital, he still keeping his shop open, and spreading the disease among his neighbours. After the death of one of his children he borrowed a suit of black clothes from a neighbour, which was sent back to the owner, and he put it on and took small-pox, and died within a fortnight."

In 1880 Dr. Wright,[3] medical officer of health for Cheltenham, writes : " During the last six years small-pox has been introduced into the town on twelve separate occasions by cases which . . . were imported from infected districts. All these patients I removed

[1] Smallpox and Fever Hospitals Commission, Minutes of Evidence, Question 1813.   [2] *Ibid.*, Question 4241.
[3] Supplement to Tenth Annual Report of Local Government Board, p. 100.

into the Delancey Hospital as soon as they were discovered, and in each case the disease never extended
beyond the house in which it first occurred."

Dr. Seaton,[1] late medical officer of health for the
Borough of Nottingham, reports in 1879 that "thirty-
seven out of sixty cases removed to hospital were first
attacks in households that contained many susceptible
children . . . and that only in one instance did a second
case occur." In 1879 scarlet fever broke out in Settle,
a rural sanitary district with about 15,000 of a population. "Three young children who were affected with
the disease were at once admitted, together with their
mother (into the Settle Infectious Hospital), and after
disinfection, both of premises and clothing, two elder
children who were the main bread-winners of the
family were allowed to resume their work at a neighbouring mill. No further spread took place, whereas
in several adjoining sanitary districts, having no means
of isolation, scarlet fever, at about the same date,
became widely prevalent."

In the Solihull rural district, with a population of
about 20,000, Dr. Page, the medical officer to the
hospital, states that with the exception of one or
perhaps two cases in which a second attack of
scarlet fever had occurred within two days of the
removal of the patient, the disease having obviously
been contracted before such removal, no spread of
that disease had taken place in any house from which
the first patient had been removed.

[1] Annual Summary for 1879, by Edward Seaton, M.O.H.

It will thus be seen, that, by means of hospital provision, and proper disinfection, the danger of infectious disease spreading among the inmates of even small houses may be reduced to the minimum, and an infected house and family, instead of being a menace to the neighbourhood for months, may be disinfected and rendered harmless in a few days. The various infectious diseases can be recognised sufficiently early by medical men ; the time the various fevers take incubating in the system after infection is fairly well known ; the period over which the danger of infecting others extends can be reckoned with comparative precision ; powerful agents and methods for destroying infection in premises and clothing have been devised. We have only therefore to take advantage of the knowledge acquired, to be able to check the spread of infectious diseases, if not to exterminate them altogether.

Until the Infectious Disease Notification Act was passed, it was not always possible to find out such cases in time to have much effect in checking the progress of an epidemic. Now in those districts where this Act has been adopted, both medical practitioners and householders are bound forthwith to notify all cases of infectious disease to the Local Authority. This Act has been adopted over a large part of this country, but in rural districts it is seldom that hospitals and ambulances are provided. Until this is done, the information obtained by means of the Infectious Disease Notification Act in checking the spread of epidemic

diseases, cannot always be used with success.   Indeed
it is impossible in some cases to do more than to watch
the disease gradually spreading. Even without hospitals
the information collected by means of this Act is of
considerable service in many cases in preventing the
spread of infection.

CHAPTER IV

THE air of a hospital ward, or of any apartment, where
cases of infectious disease are accommodated, is contaminated by gases, which issue from the bodies of the
inmates ; more particularly from the lungs by the
breath, and from the skin by perspiration, and by
invisible transpiration. It is also rendered impure by
the products of the combustion of gas, oil, candles,
etc. These gases consist of carbonic acid, carbonic
oxide, ammonia, sulphuretted hydrogen, etc. They are
always present to a certain extent, but with ordinary
ventilation and cleanliness, they are prevented from
accumulating to the degree that becomes a source of
danger.

In addition to these gases, however, other more
dangerous organic products are given off from the
bodies of the inmates, in the breath, perspiration, and
other secretions and excretions. These are the products of the tissue-waste of the body, which is more
rapid in the case of the sick. When these products
accumulate to any extent, they decompose, make the

air close, and cause a peculiar fœtid smell, which is easily noticed on entering an ill-ventilated ward or room. Any porous material used in the construction of a ward or room, such as a soft wooden flooring, or wooden or papered or plastered walls, is liable to absorb them. Here they undergo further changes, and only require moisture to give off more noxious effluvia. These products consist in part of "minute particles of solid or semi-solid insoluble matter, derived directly or indirectly from the bodies of patients of which they once formed a part:"[1] such as epithelial, or mucous, or pus cells, or minute particles of other excretions or secretions. Some of these particles, both before and after removal from the body, afford shelter and nourishment to various micro-organisms, some of which are harmless, while others are liable to cause, or aggravate disease. Wherever these particles find a lodgment, as on inequalities of walls and floors, or are absorbed by any material in a ward, there these micro-organisms remain, probably in a dormant or quiescent condition, but retaining their vitality. If, by any chance, such of them as are *pathogenic* or disease-producing, are disturbed, and so find their way into the body of any of the inmates, they are apt to multiply in his system and cause disease, or otherwise weaken him and render him less capable of recovery from any illness from which he may be suffering.

Before the nature of these organisms was so well

[1] Dr. Billings, *Hospital Construction and Management* (John Hopkins), p. 11.

known as is now the case, Sir John Simon[1] stated that "in ill-kept hospitals wounds go on badly; instead of running their normal courses of rapid recovery, they—whether accidental wounds, or wounds made by operative surgery—undergo certain characteristic morbid changes. Erysipelas will frequently attack them, so will other morbid processes akin to erysipelas, such morbid processes as those of gangrene, phagedæna, or putrefactions of effused or otherwise stagnant blood, and reopenings of half-healed arteries and veins, and septic and suppurative infections of the system, and so forth. . . . And if the hospital receives lying-in women, these infections will constitute different forms of so-called puerperal fever." "In such exhalations are embodied the most terrible powers of disease, the spreading flames, as it were, of some infections, and the explosive fuel of others, and any air in which they are let accumulate soon becomes a very atmosphere of death." Since that time specific micro-organisms have been demonstrated in, and proved to be the cause of, the various diseases or morbid processes enumerated by Sir John Simon.

Long before micro-organisms were proved to be the agents of infection, Dr. Burdon Sanderson showed that "contagium" consisted of particles, and was neither fluid nor gaseous. In a paper published in the Twelfth Report of the Medical Officer of the Privy Council in 1869, on "The Intimate Pathology of Contagium," he stated that contagium was neither

[1] *Local Government Board Report*, 1863, pp. 52 and 59.

soluble in water, nor capable, without losing its properties, of assuming the form of vapour, and that it was
not possible to admit that contagium was an insoluble
solid or liquid, without admitting at the same time that
it consists of separate particles which scarcely differ
from the animal liquids in which they are suspended,
either in their specific gravity, or in the degree they
affect the transmission of light.   He diffused vaccine
lymph, and found that it was inert.   He dissolved a
drop of lymph in a large volume of an inert liquid, and
found its action proportionally uncertain.   " If," he
states, " on the other hand, contagium were soluble it
could not be explained, for in that case each of the
10,000 drops, in which the one drop had been dissolved, would be equally active.   Assuming it to be
particulate, it follows that the myriads of particles,
which were before distribution in one drop, are
scattered through 10,000 drops ; and inasmuch as
there is nothing excepting the influence of currents, to
ensure the equal distribution of the particles, it is
clear that in some regions of the liquid, the distance
from each other will be greater, in others less.   Consequently when a trace of the liquid thus feebly impregnated with contagium is taken up on the point
of the lancet, the chance that the little drop will
or will not contain particles may be stated numeric-
ally, by the fraction which expresses the degree of
dilution."

" And here it is of importance to notice that the
same explanation applies to a fact of common observa-

tion with respect to all of those diseases which are contagious at a distance. The question is frequently asked, How does it happen that a person may be exposed every day for many months to the contagium of typhus with immunity, and yet be eventually attacked without any change whatever being made, either in his own condition, or in that of the infected media by which he is surrounded? If contagium were gaseous, the fact would be inexplicable. . . . Assuming it to be insoluble and particulate, the question of mediate contagion must, like that of direct contagion, be one of chance. Just as in the case of inoculation, the effect of dilution shows itself exclusively in the proportion of failures to the total number of insertions, so in exposure to infected air, the effect of distribution of the poison through a large volume of air shows itself in the proportion of escapes to the total number of exposures which the individual passes through. And just as in the former instance, the last inoculation of a series is just as likely to be the one by which a particle of contagium is introduced as the first, so in mediate contagion the last exposure is just as likely to be the effectual one as the first."

" With reference to their mode of action, we have examined into those considerations which seem to render it probable that they are organised beings, and that their power of producing disease is due to their organic development; and we have accepted the doctrine as the only one which affords a satisfactory explanation of the facts of infection, and in particular,

of those which tend to show that within the body of
the infected individual the particles of contagium
rapidly reproduce themselves, while out of the body
they are capable of resisting for very long periods the
influence of conditions, which if not restrained by
organic action would produce chemical decomposition."

Within the last fifteen years, great improvement
has been made in the methods by which the nature of
these micro-organisms can be investigated.   Formerly
various species of these organisms were, on account of
the defects of the methods in use, liable to get mixed
in the process of examination and cause confusion.
There was no efficient way of separating one species
from another, and of cultivating it separately so as to
arrive at any reliable conclusion as to its behaviour
under various conditions.

Pasteur opened the way for future workers by his
method of obtaining a comparatively pure culture of
the micro-organism of yeast by means of food more
suited to its nutrition than to that of other micro-
organisms usually associated with it.   This method
was improved by Klebs in 1873, by combining the
fractional method with that of providing specially
prepared food.   Sir William Roberts and Cohn a few
years later found that some species were more easily
killed by heat, and in this way obtained pure specimens
of some species of micro-organisms.   In 1878-79, Sir
Joseph Lister and Dr. Naegeli obtained pure cultures
by distributing a number of organisms in large quan-

tities of fluid, and making greater and greater dilution of these organisms in broth until the dilution was such that a single drop did not contain more than a single organism.[1] In 1877 Sir William Roberts, in an address delivered to the British Medical Association, "On Spontaneous Generation and the Doctrine of the Contagium Vivum," showed that organic matter has no inherent power of generating bacteria, and that bacteria are the actual agents of decomposition. "As to the nature of the infecting agents," he states, "we can say positively that they must consist of solid particles, otherwise they could not be separated by filtration through cotton wool."[2] Professor Tyndall has demonstrated that air which is optically pure—that is, air which is free from particles, has no fecundating power. Dr. Burdon Sanderson, writing in 1874, inferred from the persistence of the contagium of splenic fever outside the body, in stables and other places, and its short-lived and fugitive existence in the blood, that the micro-organism must have two states of existence, namely, that of the perishable bacteria found in the blood, and some other more permanent form in which like seeds or spores they were capable of surviving for an indefinite period. Professor Cohn, guided by the botanical character of the rods, came to the same conclusions. These previsions, Sir William Roberts states, were proved by the researches of Koch to be

---

[1] *Bacteria and their Products*, by Dr. German Sims Woodhead, p. 304.
[2] *Spontaneous Generation and the Doctrine of the Contagium Vivum*, by Sir William Roberts, M.D., p. 12.

perfectly exact.[1]   Koch cultivated the bacillus of
splenic fever (*bacillus anthracis*) in a drop of the aque-
ous humour of the eye, in an incubator at a temperature of about 98.4° Fahr., between two microscopic
slides, and examined the progress of events now and
then under the microscope, and he actually saw the
spores or seeds forming.   He found that mice were
peculiarly liable to splenic fever, and by experimenting
on them he found that the bacillus anthracis lived only
for a comparatively short time, but that the spores or
seeds retained their vitality for an indefinite period.
"They could be reduced to dust, wetted and dried
repeatedly, kept in putrefying liquids for weeks, and
yet at the end of four years they still displayed an
undiminished virulence."[2]

In 1881 Koch[3] published his improved methods
for the investigation of the nature of pathogenic
organisms. This consisted mainly of a process of
rendering the bacteria in fluids and animal tissues visible
by means of staining reagents, and of obtaining pure
cultivations of different species of micro-organisms by
means of solid or semi-solid media, instead of the fluid
in which bacteria were previously cultivated.   He
states :—"It being perfectly clear that efforts in this
direction are in vain, I have abandoned the principles
on which pure cultures have hitherto been conducted,

[1] Koch, "Aetologie d. Milz-Brand-Krankheit" in Cohn's *Beiträge z. Biologie d. Pflanzen*, Bd. II. Heft ii. 1876.
[2] *On the Doctrine of the Contagium Vivum*, by Sir William Roberts, M.D., p. 37.
[3] *Mittheilungen aus dem kaiserlichen Gesundheitsamte*, vol. i., Berlin, 1881. Translated by Victor Horsley, B.S., F.R.C.S., for New Sydenham Society.

and have struck out a new path, to which I was led by
a simple observation which any one can repeat. If a
boiled potato is divided, and the cut surfaces exposed
to the air for a few hours, and then placed in a moist
chamber (as, for instance, on a plate under a bell jar
lined with wet filter paper) so as to prevent drying,
there will be found by the second or third day (according to the temperature of the room) on the surface of
the potato numerous and very varied droplets, almost
all of which appear to differ from each other. A few
of these droplets are white and porcellaneous, while
others are yellow, brown, gray, and reddish ; and
while some appear like a flattened-out drop of water,
others are hemispherical or warty. All grow more or
less rapidly, and between them appears the mycelium
of the higher fungi ; later the solitary droplets become
fused together, and soon marked decomposition of the
potato occurs. If a specimen is taken from each of
the droplets, so long as they remain distinctly isolated
from each other, and is examined by drying and staining, a layer of it on a cover glass, it will be seen that
each is composed of a perfectly definite kind of micro-
organism." [1]

With few exceptions, every droplet or colony is a
pure culture, and remains so until by growth it pushes
into the territory of a neighbour ; the exceptions
being those cases where two spores fell quite close
to one another. If, on the other hand, the spores

---

[1] *Mittheilungen aus dem kaiserlichen Gesundheitsamte*, vol. i., p. 37 : Berlin,
1881. **Translated** by Victor Horsley, B.S., F.R.C.S., for New Sydenham Society.

were to fall into a nutrient fluid, the organisms endowed
with motion would have dispersed themselves rapidly
throughout the liquid, and mixed with those not so
endowed. Some bacteria, however, were found not to
grow when planted on the surface of the potato.
This difficulty, however, was surmounted by pre-
paring nutrient gelatine, to be used in a similar
manner.

By the introduction of the exact method of cultivation of bacteria on solid media and of staining bacteria,
"there has suddenly set in a flow of researches and
discoveries, which, like a snowball in rolling from the
top of a snowhill, is constantly increasing in size, and
its advent is felt in wider and wider circles."[1]   Since
that time the *bacillus tuberculosis* has been conclusively proved to be the cause of consumption.   In
1883 Koch demonstrated that this organism could be
separated from the tissues of the body, cultivated
outside the body in nutrient media, and was capable
of causing consumption when introduced into the body
of some animals, and further, that the same organisms
were found in the diseased tissues of these animals.[2]
It was also found that "tubercle bacilli taken from the
lungs of phthisical bodies, and buried for years, still
possess their characteristic reaction, and are capable
of producing tuberculosis on inoculation.   This would
mean that the tubercle bacillus, even after years, was

---

[1] Klein on "Pathology of Infectious Diseases," *Hygiene, and Public Health*, (Stevenson and Murphy), vol. ii. p. 34.

[2] *Bacteria and their Products*, G. S. Woodhead, p. 206.

still living and capable of carrying on an existence
outside the living human or animal body."[1]

The anthrax bacillus, which is the cause of woolsorters' disease in human beings, has been described
so long ago as 1849 and 1850 by Pollender, Rayer,
and Davaine, as occurring in the blood of animals
dying from splenic fever. It was left to Koch, however, to give absolute proof that this organism was the
actual cause of splenic fever. To this I have already
referred. In his Reports to the Medical Officer of
the Local Government Board in 1881 and 1882, Mr.
Spears conclusively proved that woolsorters' and rag-
sorters' disease was caused by infection from the dust
of infected wool and hides. The blood in the general
circulation, and in all the vessels of the organs of persons
dying from these diseases, contains the *bacilli anthracis.*

The bacillus of diphtheria, isolated by Löffler,
called by Klein the Klebs-Löffler bacillus, has been
conclusively proved to be the cause of diphtheria. Dr.
Woodhead states that diphtheria is entirely dependent
for its specific symptoms on the diphtheria bacillus,
experiment corroborating clinical observation in a
most remarkable manner.[2] This bacillus also lives
outside the human body. It lives in gelatine, and
grows luxuriantly in milk. By subcutaneous inoculation
it causes diphtheria in guinea pigs.[3] Cats have been

[1] "Pathology of Infectious Diseases," *Treatise on Hygiene and Public Health,*
vol. ii. p. 75 (Stevenson and Murphy).
[2] Address on Bacteriology, delivered before the British Medical Association
at Nottingham, 1892.
[3] "Pathology of Infectious Diseases," *Treatise on Hygiene and Public Health,*
vol. ii. p. 156 (Stevenson and Murphy).

F

known to contract the disease. "In houses where human diphtheria obtained, cats have been known to become ill either antecedently, or coincidently, or subsequently: they appear to have some kind of throat illness and cannot swallow; they sneeze and their eyes water: as a rule bronchial mischief is noticed early, and if the disease is protracted through several weeks, as it sometimes is, they become much emaciated and die."[1] Klein also states that such a disease exists naturally among cats, and in one case he had seen such a cat after several weeks' illness, showing paralysis of the hind extremities, probably the same as diphtheritic paralysis in a human being. He had the opportunity of examining sections through diseased portions of the lungs of two cats dead of such a disease, and he found diphtheria bacilli in considerable numbers. He also gives very important evidence in favour of the belief that cows suffer from diphtheria, and communicate the disease in the milk to human beings.

The strepto-coccus scarlatinæ has been found to be associated with scarlet fever in human beings. Klein has been able to produce a definite eruptive disease in six milch cows, on the skin generally, and on the teats and udders particularly, by inoculating them with the human strepto-coccus scarlatinæ. Cows suffering from a similar disease have by their milk communicated scarlet fever to human beings.

Typhoid bacillus.— Gaffky[2] is the first to have

---

[1] Dr. Klein in Report of Medical Officer, Local Government Board, 1889.
[2] Mittheil. aus dem kais. Gesundheitsamte, vol. ii. p. 372.

successfully isolated and cultivated this bacillus, shown by Eberth to be found in the swollen mesenteric glands of persons who have died of typhoid fever. This bacillus grows on gelatine, in alkaline broth, in blood serum, and on boiled potatoes and other media. Klein is not satisfied that this is the bacillus of typhoid fever. He, however, states that [1] "the virus must be a microbe which evidently can live and thrive in sewage and on the soil; further, that it can maintain its vitality even when placed in such a poor medium as water; it must be possessed of considerable resistance towards putrefaction, towards cold, and towards the gastric juice; when brought in the stools of a typhoid fever case into drains, cesspools, and the like, it must be capable of preserving its life for an indefinite time. All these are requirements postulated by what is known of the epidemiological relations of typhoid fever."

Micro-organisms have also been demonstrated to be the cause of cholera, epidemic pneumonia, relapsing fever, erysipelas, septicæmia, tetanus, and some diseases of animals. On the other hand, in typhus fever, whooping cough, measles, varicella, and oriental plague, the existence of pathogenic organisms is as yet a matter of pure inference. Klein, however, states in the exhaustive contribution to the pathology of infectious diseases already referred to, that "the evidence that infectious diseases owe their origin to specific, well characterised micro-organisms is no longer a

---

[1] "Pathology of Infectious Diseases," *Treatise on Hygiene and Public Health*, vol. ii. p. 167 (Stevenson and Murphy).

matter of pure assumption, but one of exact science.
. . . Just as a chemist is enabled to detect and to
demonstrate with tolerable exactness the existence in
special cases of a chemical poison,—*e.g.* arsenic, phos-
phorus, strychnine, etc.—the material that has caused
disease and death, so nowadays can the pathologist by
microscopical, cultural, and animal experiments demonstrate the specific *materies morbi.*"

"The study of micro-organisms connected with
specific disease, *i.e.* their contagia, their nature, habits,
and power of growth outside and within an infected
individual, the manner in which they leave, and in
which they enter, the human and animal organism, the
nature of the changes which they induce in the material
in which they grow and multiply, be that material the
human or animal organism, or organic matter outside
these—all this is now well understood."

The micro-organisms of many diseases have thus
been proved by bacteriologists to live and multiply in
various media outside the human body for varying
periods, and to produce the same disease if, by inoculation or otherwise, they gain access into the body of an
animal or a human being. They have also proved
that some micro-organisms have the power of forming
spores or seeds infinitely more hardy than themselves
(just as the ripe seeds are more hardy than the parent
plant), and that these spores, after apparently any
amount of ill usage, retain their vitality for years, and
are after all capable of germinating, and thereby re-
producing disease.

It may now be shown how the experience of bacteriologists in this respect coincides with that of other observers in the field of preventive medicine.

Surgeon-General Billings,[1] U.S.A., states that he "had seen scarlatina produced by some of these particles which had been preserved in a blanket carefully packed away for years."

The following case is quoted from the *Boston Post*:[2]—"A large picture book had been used by a boy during his illness with scarlet fever in 1846. The book was packed away in a trunk for twenty-six years. Finally it was brought to England and a child two years old became its possessor. A fortnight after receiving it he was attacked with scarlet fever, and to the doctors attending the case there appeared no other means by which the child could have been infected."

Dr. Carpenter, in a paper read before the British Medical Association in 1875, records a case[3] in which "a proprietor ordered the removal of some old houses which had lain in a state of ruin and uninhabitable for many years. Eight men were employed in digging the walls, and every one was attacked with smallpox, the germs of that disease having lain in these houses all those years."

Dr. Thorne Thorne[4] states that instances have come under his own notice in which the facts warranted

[1] *Hospital Construction and Management*, p. 12 (John Hopkins).
[2] *Epidemics, etc.*, p. 264, by the Hon. Rollo Russell.
[3] This case is believed to have occurred in the Island of Mull : Dr. Carpenter got the information second-hand from a practitioner in Oban.
[4] *Transactions of the Epidemiological Society*, vol. iv. pt. 2, p. 234.

the conclusion that the poison of diphtheria had been retained for months about premises in which cases of this disease had previously occurred.

He also states in regard to enteric fever: "I know a detached house which stands in large grounds in a country district, and which was occupied by a groom and his family, amongst whom enteric fever prevailed in the autumn of 1872, one case being fatal. This family continued to occupy the house for nearly two years after this occurrence, but they left it some time in 1874, in consequence of the departure of the owner of the estate on which it stood. From that time the house remained unoccupied until February 1876, when it was retenanted by new inmates, and exactly within fourteen days of these latter taking up their abode there, enteric fever broke out amongst them, and a most careful inquiry led both the medical man in attendance and myself to the conclusion that the disease was not imported. I also know an isolated parsonage in which the families of two successive vicars have been attacked with enteric fever." He also states that Dr. Ogle, in some of his reports on the sanitary condition of East Herts district, refers to cases in which attacks of this disease have in successive autumns occurred in isolated dwellings where, so far as could be ascertained by the strictest investigation, no new source of infection could have come into operation.

Dr. Klein states that "it is also known that a room in which a diphtheria case has once existed, may for years harbour the contagium of diphtheria, so that any

new comer or inhabitant may contract the disease ;
moreover, it is known that in a locality in which
diphtheria has once been rife, the disease may at any
time re-appear, and in these instances the transmission
of the contagium from sewers is maintained by some
sanitarians."

I have already referred to an isolated farmhouse in
which outbreaks of typhoid fever occurred in 1889,
1890, and 1892. From a careful consideration of the
facts of this case, I am convinced that the infection
remained about the house during that time. There
was another case of an isolated crofter's house separated
by about one mile of uneven hilly land from any other
dwelling. In 1881 five of the inmates of this house
suffered from typhoid fever, one of whom died in 1882.
In 1883 another inmate died of the same disease. In
1884 another death occurred in the house, the cause of
death being registered as "perforation of the bowels,"
which is a frequent occurrence in typhoid fever. In
1887 another of the inmates died from diarrhœa, which
is also a symptom of typhoid. In the autumn of 1889
two boys from a town distant about fourteen miles
lived in this house for a month. They returned home
ill with this fever. At the time of the first outbreak
in 1881 several families in a neighbouring township
are stated to have suffered from typhoid fever, and the
infection is believed to have been conveyed into the
house from them. Since that time there is no history
of the disease in the neighbourhood of this house, and
a careful inquiry failed to discover any way by which

it was again introduced into the family. There was
no W.C. attached to this house. It was without any
sewers or drains. The water supply was pure. The
walls were damp and mildewed. The floor in one
apartment consisted of mud, in another, of damp,
rotting boards. The ground against the back wall
was considerably above the level of the floor. The
walls were built of rough unhewn stones, and the rain
water trickled down the outside from the thatched roof.
The soil about the house was more or less saturated
with organic matter, animal and vegetable. I know
of two other houses where typhoid fever broke out
more than once, under almost similar circumstances.

Dr. Maxwell Ross, county medical officer for
Dumfriesshire, in his first Annual Report, states :
" There were two cases [of typhoid fever] in
St. Mungo, one of them in a house in which the
germs of fever seemed to have lurked for years, it
having become one of the plague spots of the
county."

It will thus be seen that the facts proved by
bacteriologists regarding the ability of the organisms
of infectious disease to live outside the human body
and to retain their pathogenic power, exactly agree
with the experience of observers who noted the
behaviour of epidemic disease.

There appears to be some evidence in favour of
the conclusion, that the organism or agent of infection,
in the case of some diseases, is less capable of
retaining its vitality or power of infection outside

the human body, than in the case of some other
diseases. I know myself, for example, of an island
on the west coast of Scotland where whooping-
cough was introduced about twenty years ago. A
large number of the population suffered from the
disease. There was no system of disinfection followed.
Many of the houses afforded innumerable lurking-
places for the organisms of infection, yet children
were born and families reared in these houses and no
one suffered from whooping-cough until last year,
when the disease was again introduced by visitors
and spread rapidly. I have also never been able
to trace an outbreak of measles to infection remaining
in or about dwellings. I have seen it spread until,
as it were, the fuel was exhausted, and then dis-
appear for many years, until the infection was again
introduced.

There seems therefore to be some grounds for the
belief that the organisms of these diseases may not
be able to live in the same media as those of other
diseases, and the fact that bacteriologists have not yet
been able to isolate them favours this view.

Such are the special impurities or sources of
danger to be guarded against in the isolation of
infectious disease, and more particularly in the construction and management of Isolation Hospitals.
Organisms lurk in inequalities of floors and crevices
of walls. They are also liable to be absorbed by the
bedclothes, bedding, and other articles of furniture.
If the ventilation is defective, or cleanliness neglected,

the air gets contaminated and disease spreads from
the patients to the attendants, or from one patient to
another. If a hospital is properly constructed and
designed, and ".if proper ward discipline exists, no
patient is allowed to expose himself to the risk of
close intercourse with cases of contagious diseases,
and, subject to that condition, most morbid poisons are
of little effect." . . . "Contagion which will not
spread except by inoculation, or by the kindred agency
of dirty bedding or dirty towels, or dirty sponges, or
dirty fingers, or by the drinking of polluted water,
or by the effluvia from drains or cesspools, ought
to be absolutely incommunicable in hospitals. So
ought those to be where pus, or matters like pus, are
abundantly floating in the air."[1]   Drs. Thomson and
Rainy made an examination of the air in hospital
wards in 1854, and found floating in the air of one
of the wards of St. Thomas's Hospital, minute hairs,
particles of smoke, epidermic scales, vegetable fibres,
starch granules, as well as living forms, vegetable and
animal, such as vibriones and the mycelia of fungi.
In 1861, M. Chalvet of St. Louis Hospital, found a
large quantity of floating inorganic matter in the
wards of that hospital.   M. Kullman made an analysis
of the wall plaster of the walls and found that 46 per
cent was organic matter.   Experiments were made
by M. Broca,[2] which showed the presence of micro-
cocci and other bacteria, of pus globules, and of spores

[1] Report to Local Government Board, Sir John Simon, 1863, p. 56.
[2] *Revue Méd. de l'est, Revue de Thérap.*, Paris, 1874.

of epiphytes and other organisms in the air, and on and within the material of the walls of wards that have been long used.[1]

Carnelley[2] found in clean one-roomed houses 180 bacteria per 10 litres [of air]; in dirty houses, 410; in very dirty, 930; in naturally ventilated schools, 1250; in mechanically ventilated schools, 300.

A patient who has undergone an operation, or who has much broken skin, is specially liable to be infected by micro - organisms from any of the surrounding media. So much is this the case that the progress of surgical cases in a ward may be looked upon as a sensitive test of its healthiness. This is owing to the fact that in addition to the danger, present in all hospitals, of the agents of disease gaining access into the body of a patient by the respiratory system and alimentary canal, there is here another mode of access for micro-organisms due to lesions of the skin resulting from operations and various diseases. The danger has of recent years been much lessened by the antiseptic method of treatment initiated and success-fully practised by Sir Joseph Lister.

With reference to the value of this method of treating wounds in unhealthy hospitals, Dr. Burdon Sanderson writes as follows of Halle: "It is im-possible to conceive a more favourable locality for making an experiment of such a nature; first, because of the extraordinary rapid progress of the manufactures

[1] Sir Douglas Galton.
[2] "The Carbonic Acid, Organic Matter, and Micro-organisms in Air, more especially of Dwellings and Schools," *Trans. of Roy. Soc. of London*, 1887.

of the town, which has rendered the hospital too small
for its requirements, and consequently the proportion of
severe cases (as I was able to observe on going round
the wards) was very large ; and, secondly, because the
hospital itself is, not perhaps *the* worst, but *one* of the
worst that I have ever seen. Situated in the very centre
of the town, overshadowed by a huge ecclesiastical
building, and having for wards low rooms, each of
which communicates with a latrine and has its beds so
close that there is scarcely room to pass between them,
the hospital presents every circumstance likely to lead
to the development of the worst traumatic affections.
It is in such wards as these that compound fractures
and resections have been treated with a success which
is up to the best results of London surgery.   In a
word, the influence of the most unfavourable conditions
which can be conceived (the disastrous effects of
which were before so overpowering that to Prof.
Volkmann his clinical work had become a burden
instead of a joy), have been simply neutralised by the
conscientious carrying out in all its details, with the
earnest co-operation of the whole staff of the hospital,
of the Edinburgh method." [1]

The antiseptic method can thus be shown to be
singularly successful in preventing the micro-organisms
of various diseases from gaining access into the human
system through wounds and other lesions of the skin.
There is no similar specific method to prevent the

[1] *The Infective Processes of Disease.* Lecture delivered in the theatre of the
University of London, by Dr. Burdon Saunderson.

micro-organisms of the ordinary infectious diseases
from entering the system, in the breath or in food or
water. It is therefore all the more necessary that in
Isolation Hospitals the wards should be so constructed
as to allow almost an unlimited supply of fresh air, and
ample room for each patient, and that the material used
in their construction and furnishing be such as not
to retain or absorb the organisms of infection. This
is particularly necessary in small hospitals, the wards of
which may be used at one time for one infectious
disease, and at other times for other infectious
diseases.

CHAPTER V

THE ESTABLISHMENT AND ERECTION OF ISOLATION HOSPITALS—
PRELIMINARY

THE expense of erecting and maintaining Isolation Hospitals may at first sight appear to be out of proportion to the small number of patients *treated* in them. The primary object of providing Isolation Hospitals is, however, to *prevent infection from spreading* by the seclusion or separation of persons suffering from infectious ailments from all who may be susceptible to catch these diseases.

A few instances may make this plain. The Delancey Hospital, Cheltenham, was erected and furnished at an expenditure of £11,121 : 2 : 0. The total number of beds was 32. During the two years 1879 and 1880 the total cost of maintenance, excluding certain items for additional furnishings during the second year, was £734 : 13 : 2 and £731 : 17 : 7, each of these amounts including a sum of about £450 for wages and housekeeping, and an honorarium of £42 to the medical officer. During these two years 7 cases of smallpox and 53 cases of scarlet fever were isolated in the hospital. The gross cost of maintenance was thus at the rate of £24 : 8 : 10 per patient. The simple treatment and maintenance of these 60 cases in the hospital was, however, only a small item of the benefit that accrued. In six years, by means of the hospital, small-pox was prevented from spreading among the community on twelve separate occasions, and the same testimony is borne as regards the isolation of scarlet fever. Although the cost for maintenance at the above rate looks very large for the simple treatment of sixty cases, it is very small in comparison to the loss that would result from epidemics of smallpox and scarlet fever among the 43,000 people whom the hospital served.[1]

At Settle a hospital for ten beds was provided in 1878 by the Local Authority at an expense of £1061 : 2 : 6, devoted to the improvement of an old hospital handed over to them by the Board of Guardians. The hospital had, according to Dr. Thorne Thorne, been but little used until he visited it. During the second quarter of 1879, three children affected with scarlet fever were admitted, together with their mother. No spread of the fever took place, although widely prevalent in surrounding districts. The simple treatment of this woman and her three children was in itself of very little good compared with the prevention of an epidemic of the disease in the district. It was much more economical for the population to provide for them in a well-appointed hospital than to incur all the expense, misery, and death which might have resulted had the epidemic

[1] Supplement to Tenth Report of Medical Officer, **Local Government Board**, pp. 100-102, reissue 1893.

spread among them. Other instances like the above may be found by reference to Dr. Thorne Thorne's Report.[1]

Wherever a hospital for infectious disease is provided, immediate advantage should be taken of it as soon as a single case of infectious disease appears. The case should be at once removed to it and isolated. For this purpose the hospital must be always kept in perfect readiness for the admission of patients. Every hour's delay increases the risk of the spread of infection. If the patients are not isolated until cases multiply and the disease gets a footing, the hospital will be of little service for the prevention of an epidemic. The disease will spread in spite of what is done, and as far as the primary object of preventing disease is concerned, the hospital will be an encumbrance and expense. The cost of maintaining the hospital will be proportionately less, and the benefit resulting from it will be proportionately greater, according to the perfection of the facilities provided for the immediate admission of patients to it, and the advantage taken of these facilities by the public.

The number of beds that should be provided in an Isolation Hospital for a small town or a rural district, and the provision that should be made for the isolation of separate infectious diseases in the same hospital at the same time, is rather difficult to determine.

In large towns the population is crowded, and one well-equipped hospital of large size, and constructed

---

[1] Supplement to Tenth Report of Medical Officer of the Local Government Board, reissued 1893.

in such a way as to isolate different infectious diseases at one and the same time, may be erected within easy distance of the community. In rural districts this is out of the question. In large towns it has been stated that about 1 bed per 1000 of the population would be sufficient provision for the isolation of infectious disease to meet ordinary emergencies. This, however, will not hold good in the case of rural districts or even small towns.

The number of beds necessary will be found to vary in different localities. In some places there is more danger of infectious disease being frequently introduced by visitors and otherwise, and of the outbreak of more than one disease at the same time. The number of beds in the hospital should also to a certain extent depend on the character of the house accommodation in the district. If the houses of many of the inhabitants are commodious and well isolated, and so constructed that cases of infectious disease may be treated at home without danger to the inmates or to the public, there is not the same necessity for a large number of beds in the hospital. The opposite is the case if the houses are small and crowded. In localities remote from large centres of population the visitations of infectious diseases are as a rule rare, and it is seldom that more than one disease breaks out at the same time. In such localities the spread of infection is also more slow than in towns, owing to the greater distances between houses, and the less frequent communication between

G

families.   The spread of infection is, however, more
likely to get a certain headway in such places.   People
are often taken unawares, and medical attendance is
not so convenient as in more populous centres.   Several
members of a family are often taken ill before the
disease is suspected to be of an infectious nature.
This must be considered in reckoning the number of
beds that should be provided in thinly-populated rural
districts.   If it were possible to convey patients to a
hospital from very long distances in rural districts, it
would probably be found that the number of beds
required would be less than in towns.   This, however,
is not the case.   However small a population may be,
the number likely to be infected before the disease is
discovered must be taken into consideration.   This is
probably the reason why the medical department of
the Local Government Board recommend that at least
four beds should be provided for villages.

The distance that patients suffering from infectious
diseases may be conveyed with safety to a hospital is
a matter for serious consideration.   On this depends
the number of hospitals that should be erected to
meet the requirements of a population scattered over
a given area.   Some cases suffering from infectious
diseases cannot, without running great danger, be
removed to a hospital, however near.   Such cases,
however, are exceptional.   The great majority of
cases may be conveyed considerable distances during
the first few days of their illness without any danger.
Even in cases that cannot be removed at once it may

be possible to have them conveyed to a hospital at a later stage with safety.

The following evidence given before the Infectious Hospitals Commission may be of interest as bearing on the distance that patients may be conveyed to hospitals. Dr. Murphy,[1] Medical Officer of Health for the London County Council, is of opinion that although many severe cases were conveyed to the London Fever Hospital a distance of five or six miles, they were not aggravated by the conveyance.

Dr. Collie,[2] late Medical Superintendent to the Homerton Fever Hospital, states that nearly every case of smallpox might be removed twenty miles as safely as one; and that taking it broadly, the removal of smallpox patients is not injurious to them, inasmuch as after carefully examining 6771 cases, he found that out of the total number only one was found dead on arrival, and 75 died after 24 hours' residence in the hospital, 26 of these being of a malignant type (and therefore would have died under any circumstances), 21 confluent, and 2 semiconfluent. They were all brought in old cabs. He would say that 16 or 17 miles would not hurt, and that in properly constructed ambulances, the worst cases might be brought from their homes to the hospital, under proper superintendence, without hastening death. He had frequently known cases conveyed to the hospital seven or eight miles, and some of them very serious cases.

---

[1] Smallpox and Fever Hospitals Commission, Minutes of Evidence, Questions 3021 and 3024. [2] Question 2119.

Dr. Thorne Thorne,[1] Chief Medical Officer to the Local Government Board, stated that his experience covered a few isolated instances where patients were conveyed a distance of from eight to ten miles, and many instances a distance of from four to five miles, and that he had not known any evil effects following on the removal, except where the patient was practically dying before removal.

Dr. Bernard,[2] late of the Stockwell Hospital, however, states that in a bad case of smallpox, after the disease has shown itself some thirty-six hours, a very grave risk is run by removal.

Dr. Littlejohn,[3] Consulting Medical Officer for the Board of Supervision, states that, generally speaking, for such a district as the three Lothians and the adjacent counties, the distance should never be more than six miles, care being taken in all cases to provide a suitable attendant, and warmth in the shape of hot-water bottles, together with restoratives.

Dr. Gayton, Medical Superintendent of the North Western District Hospital of the Metropolitan Asylums District Board, after twenty-three years' experience in the hospital treatment of infectious diseases, writes to me as follows :—" In smallpox, evidence, I think, exists that removal, practically to any distance, after the rash is well out and therefore the temperature down, is a harmless proceeding, but to do so during the premoni-

---

[1] Smallpox and Fever Hospitals Commission, Minutes of Evidence, Questions 1072 and 1073.   [2] Questions 3177 and 3179.
[3] Circular, Public Health, No. VI., 1889.

tory fever, with a temperature of 104° or 105°, has always appeared to me risky, as also about the eighth or ninth day, when the secondary fever has set in. Of scarlet fever removals, I feel the shorter the distance the better the chance of recovery, and also that in those cases in which the rash is 'livid red,' as Fagge puts it, it would be of immense advantage to the patient not to expose him at all to cold air, such as he might encounter in the winter months. Shortly put, my feeling would go to treat cases where they arose, and draw a cordon around them, rather than send them for an excursion with temperatures by which the heart's action is jeopardised. Such, however, being impossible, the only rule to lay down is to make the journey as short as possible. Enteric speaks for itself, and says with no uncertain voice, Leave me where I am or move me with the greatest care and the shortest distance possible. I have seen cases perforate within an hour after admission, presumably due to the jolting of an ambulance, and to the unscientific handling, the journey not having exceeded one mile. In reference to diphtheria, I see very little objection to their removal to hospital and to some distance, if the transfer is made early and the horizontal position is maintained, but to do as is common now, admit to-night and certify the death to-morrow, is clearly a mistake, and only serves to credit us with a mortality with which we have nothing to do."

Dr. Watt, Medical Officer of Health for the County of Aberdeen, states that patients have been conveyed

in that county a distance of eleven to fourteen miles
without any ill effects. I have myself known several
cases of scarlet, typhus, and typhoid fevers having been
conveyed distances varying from eight to twelve miles
without any dangerous result.

It will thus be apparent that no hard and fast rule
can be applied to the conveyance of patients suffering
from infectious diseases. Every case must be decided
on its merits. Not only must the condition of the
patient at the time be considered, but the means of
conveyance, the state of the roads and weather. It is,
however, probable that the great majority of patients
suffering from any of the common infectious diseases
prevailing in this country may be conveyed, in proper
ambulances and with careful attendance, for considerable distances without running any risk.

Dr. Thorne Thorne [1] recommends that in rural districts the hospital should be erected not more than four
or five miles from the more populous portions of the
districts concerned, and in the case of towns not more
than two miles. Such a hospital would probably
serve for the isolation of patients from less populous
places over an area distant by road twelve to fifteen
miles. To do this, however, a well constructed carriage
should be provided, and the removal effected at an
early stage of the disease, or if this could not be done
with safety at the time, it might be accomplished after
the patient recovered sufficiently to bear the journey.

[1] Supplement to Report of Medical Officer of Local Government Board,
1880, p. 8.

Outbreaks of infectious diseases almost invariably start in the more populous localities. The disease, if unchecked, afterwards spreads to the outlying parts. A hospital would be of more service in preventing the spread of disease to the outlying districts than in isolating cases after they had occurred there.

The feeling of the population should also to a certain extent be considered in settling the distance patients are to be conveyed to a hospital. There is in some places a feeling among many against sending their friends to a distance. This might militate against the use of the hospital afterwards. The nearer a hospital is provided, the greater the chance of its proving beneficial. To secure this, small burghs dotted over the country should combine with the Rural Authorities for hospital provision. Under present circumstances this, however, is difficult to accomplish. "Local[1] jealousies, the fear of expense, and other causes have interfered with the extension of this plan, but the appointment of County Councils seems to us to afford an opportunity of removing the difficulty. Except in London the County Councils have not yet been invested with many sanitary functions. Parliament might well increase their powers in at least one direction, by constituting them the hospital providing authorities for the whole of their respective counties. In this way the Councils would be placed on somewhat the same footing as the Metropolitan Asylums Board

[1] *Hospitals and Asylums of the World*, Burdett, vol. iii. p. 109.

of London, and probably with equally satisfactory
results." County Councils might with advantage be
vested even with greater powers in this direction than
Mr. Burdett proposes. In dealing with infectious
diseases, the wider the area under one Local Authority
and the more uniform the system adopted for the prevention of disease, if carried out with energy, the greater
is the chance of success. Under the present system
counties in Scotland are divided into districts; each
district is managed by a separate Local Authority or
District Committee, almost independent; and dotted
here and there in the districts, burghs are to be found,
entirely independent of County Councils. If one or more
of these districts or burghs do not provide sufficient
means to prevent the spread of disease, the surrounding
area will be continually infected by them. If, however, County Councils were acting Local Authorities
for the prevention of the spread of infection within
their respective counties, a uniform method would be
adopted. The stronger the public body having control of this matter, the greater the chance of measures
being carried out with efficiency. Perhaps in time the
public may come to see that even a stronger board and
further centralisation is desirable.[1]

When it is decided to build a hospital within a
certain area, the site has to be selected. If it is of
great importance that healthy sites should be chosen

---

[1] In the practice we have not much to learn, but in the policy a great deal, in
the concentration of the function, in the guidance of the practice, so that there
may be uniform and united action over the entire area of what ought to be in
reality as in name the community.—Dr. J. B. Russell.

for dwelling houses intended to be occupied by the strong and vigorous, it is of still greater importance that the site for a hospital should be healthy, as it is intended to be occupied by the sick. Persons in ill health are more susceptible to all deleterious influences than the healthy. The natural configuration of the land, the geological formation, the soil, the underground and surface drainage, as well as the relative height, and the exposure to sun, rain, and wind, should all be considered. A gentle slope at a moderate elevation is preferable to the base of a hill, a deep valley, or a level plain. Dry gravelly or sandy soil is preferable to stiff soils, such as peat or clay, likely to retain moisture. Estuary shores, river banks, marshes, made soil, or ground likely to contain organic matter, should be avoided. A hospital site should have a good exposure to the sun, and if possible be sheltered from unhealthy and cold winds. It is also desirable to have the site in such a locality that the prevailing winds do not blow towards it from any unhealthy area, or from works which would be likely to have any ill effect on the health of the inmates of the hospital. It should be within reach of a plentiful supply of pure water, and have reasonable facilities for drainage. "All these elements are of importance, every one in its place. It is obviously of no use to build a hospital in the best air in the world if neither patients nor medical officers can get to it. It is only in applying common sense to such a question, and by always giving a preponderance to conditions of highest importance, namely, pure air [and water] and space, when the other considerations can be at the same time reasonably obtained, that the best will be done for the sick."[1]

The next question for consideration is the area of a site for a hospital with a certain accommodation and of a given size. This is to a certain extent governed by the width of the space that should exist between the hospital and the area wall, and by the design or plan of the building to be erected on the site.

In the case of an infectious hospital it is necessary, in order to prevent the risk of infection from spreading to the public outside the grounds, to have a certain space between the buildings and the area wall.

Unless a reasonable space or sanitary zone is provided, infection may be communicated to the public by means of infected articles thrown over the wall, or by currents of air carrying infected dust. What might be a safe distance may be inferred from experience gained in the case of various infectious hospitals.

In regard to the London Fever Hospital, "the occupants of houses in Theberton Street, West, and of Nos. 11, 12, and 13 Gibson Square, have lived within a distance of $78\frac{1}{2}$ to 84 feet of wards occupied by typhus fever from 1863 to 1867, and by scarlet fever from 1871 to 1880. The inhabitants of No. 4, 5, 6, and 7 Barford Street have lived within 55 feet of wards occupied by typhus from 1863 to 1867 inclusive, and by scarlet

---

[1] Florence Nightingale, *Notes on Hospitals*, p. 29.

fever from 1871 to 1880. The inhabitants of Nos. 11 to 16 Barford Street have lived within a distance of from 49 to 60 feet of wards occupied by typhus from 1864 to 1868 inclusive, and by relapsing fever from 1869 to 1871. From the houses in Barford Street project W.C.'s, which reduce the distance between the houses and the opposite wards to a distance of between 36 and 50 feet. The shortest distance between a house in Charles Street and the opposite ward occupied by typhus and relapsing fever being 29 feet, while, if the garden boundary be taken to represent the nearest approach of the inhabitants to the wards, the distance is further reduced to 22 feet.

" The number of cases treated during thirty years was 18,073 typhus. As the result of my inquiry not a single case of typhus could be heard of in the houses surrounding the hospital.

" The total number of relapsing fever treated was 2078, but no evidence could be found of a single case of the disease having existed in any of the houses.

" With typhoid 6645 were treated during the thirty years, and only five cases could be heard of in any of the houses.

" As to scarlet fever, wards which contained for years together as many as twenty-eight cases at a time, have existed within 55 feet of inhabited houses, without any case of this disease occurring in these dwellings. In all the houses (ninety-six) surrounding the hospital thirty cases existed within the memory of any of the occupants. Dr. Murphy believed that thirty cases was not in excess of what might be found in any district during the same number of years with a similar class of property in a neighbourhood where a fever hospital did not exist." [1]

At Newcastle-upon-Tyne [2] the hospital was within 34 feet of a poor and crowded neighbourhood, yet although scarlet, typhus, and enteric fevers were treated in the hospital, none of these diseases spread to the outside public, and what is very remarkable, although two elementary schools were one within 45 feet and the other within 100 feet, the children attending them were not infected from the hospital by any of the diseases treated there.

Mrs. Gladstone's Convalescent Home for Smallpox Patients, according to the late Dr. Tripe, Medical Officer of Health for Hackney, was used for acute cases, and was within 94 feet of a row of houses at the back. The windows at the back were kept shut. No case of smallpox whatever occurred during the time the hospital was carried on, in the row of houses referred to. [3]

"The City of London Workhouse, which overlooks the Homerton Fever and Smallpox Hospital, and is distant from it, window to window, only 90 feet, had scarcely any case in the epidemics of 1871 and 1877, when the disease was extremely prevalent in the sur-

---

[1] Smallpox and Fever Hospitals Commission, Minutes of Evidence, Questions 3008-3013.

[2] Supplement to Report of Medical Officer, Local Government Board, 1880-81, pp. 201 and 202.

[3] Smallpox and Fever Hospitals Commission, Minutes of Evidence, Question 576.

rounding streets, although at the time the inmates were not protected by revaccination." [1]

After, however, taking all the evidence collected into consideration, the Infectious Hospital Commission states [2]—" But we feel, so long as it is not proved that 'personal communication' is adequate to the whole spread of smallpox, and so long as distant 'atmospheric dissemination' is not shown to be in the highest degree improbable, so long is it essential that in the construction and management of smallpox hospitals both sources of danger should be with the utmost care guarded against."

After taking the history of many hospitals into consideration, in so far as the spread of disease from them is concerned, the Medical Department of the Local Government Board have come to the conclusion that no part of the wards or outhouses of an infectious hospital should be nearer than 40 feet to the area wall. This, however, does not include smallpox. A wider area would probably be necessary for a smallpox hospital. From the above facts it is clear that there is no necessity whatsoever for the erection of Isolation Hospitals at such inconvenient distances from habitations as is frequently the case. In rural districts and small burghs, however, where land is plentiful and comparatively cheap, it is better to be on the safe side by building a hospital on a roomy site, reasonably isolated, but at the same time at a convenient distance

[1] Report of Smallpox and Fever Hospitals Commission, 1882, p. 26.
[2] Ibid. p. 27.

by a good road from the most populous part of the
district to be served by it.

In addition to the sanitary zone or neutral ground
all round the hospital buildings (which should not be
entered by the patients on one side nor by the public
on the other), the hospital buildings take up some room.

The amount of ground covered by the buildings
must necessarily depend on the size of the hospital—
that is, the accommodation to be provided in it.   It
also depends on the design, and the design or plan
depends to a large extent on the number of infectious
diseases to be treated at one time in the hospital.   A
small cottage administering to a couple of wards covers
much less ground than a large administrative block
serving many wards.   A number of wards in a many-
storied building covers much less ground than the
same number in one-storied blocks.

If more than one infectious disease is to be treated
at one time in the hospital, it is quite as important to
have a safe space or sanitary zone between the wards
occupied by different infectious diseases as it is to
have a safe sanitary zone between the hospital and
the public.   Indeed it is in some respects of more
importance.   A patient suffering from typhoid fever
is quite as liable to catch scarlet fever as a healthy
person, and in his weakened condition he is more
likely to die if he gets it.   As I have, however,
previously endeavoured to show, it appears that some
infectious diseases are more liable to spread to persons
at longer distances than others.   The infection of

typhus fever appears to travel by means of the atmosphere in a well ventilated ward but a short distance from the patient. The same is true of diphtheria and typhoid fever. Smallpox and scarlet fever, however, travel farther. If the same wards were to be always used for the same diseases only, they might be safely placed at varying distances apart, according to the distance infection might travel from them. This would necessitate a large number of wards, and in some cases would be unnecessarily extravagant. By erecting one or two blocks with two wards each, so far apart that the most infectious of the ordinary fevers, as well as the least infectious, may, as circumstances require, be isolated in them, this difficulty is to a certain extent overcome. It is clear, therefore, that if a sanitary zone of forty feet is necessary between the wards and the public, there should be at least an equal distance between these ward blocks and those occupied by persons suffering from other infectious diseases. In the Leamington hospital, see plan, appendix, the most infectious diseases, *i.e.* those diseases likely to spread to a distance, could at different times be isolated in the ward block on one side, while other diseases, such as typhus and typhoid or diphtheria might be treated in different parts of the isolation block on the other side of the hospital grounds.

In addition to this necessary zone some space must be allowed, such as airing courts for convalescents, drying green, as well as space required for the general working of the institution.

The size of an Isolation Hospital may appear out of proportion to the population and the number of patients treated in it. This is more particularly the case in rural districts. To any one who has not carefully studied the question it may appear unreasonable that a hospital with several wards should be built apart on the pavilion style, say for the accommodation of twenty-four patients, when from one end of the year to the other there may not be more than twenty-four patients under treatment in the whole building, and at any one time perhaps not more than half a dozen.

Yet when it is considered that in such institutions not only must there be provision for the separation of the sexes, but also for at least patients suffering from two different infectious diseases, as well as for the accommodation of nurses attending to the different diseases, the reason for what may appear to be excessive accommodation becomes apparent. Not only is it necessary to have separate wards for males and females, suffering from say two infectious diseases, but it is always desirable to have a room in which doubtful cases may be isolated from the other inmates until the nature of the illness is beyond doubt.[1] It is a very serious matter for a patient suffering from measles to be put by mistake into a smallpox or

[1] In 1890 in London 85 cases of various diseases were certified as scarlet fever, 74 as diphtheria, 164 as enteric fever, 5 as continued fever, 1 as relapsing fever, 2 as smallpox, 1 as puerperal fever, 4 as typhus fever, in all 336. All these were admitted into the hospitals of the Metropolitan Asylums Board, but were found to be suffering not from the diseases certified but from various other diseases mistaken for them.—*Stat. Com. Rep. M.A.B.*, 1890, pp. 76-80.

scarlet fever ward, as he may contract either of these diseases immediately afterwards, and while in a weakened state is more liable to die of it.

In large well ventilated and well managed hospitals, it has been found practicable on many occasions to treat a few cases of typhus, or typhoid fever, or diphtheria, in a general ward with non-infectious cases. If this be done the greatest care is necessary to prevent such diseases from spreading, and even in the best hospitals, and with the greatest care, the system failed, and is now almost universally condemned. It would probably be impossible to prevent smallpox, measles, or scarlet fever from spreading, if treated in a ward containing persons susceptible to these diseases.

In a Report to the Privy Council in 1864, by Drs. Bristowe and Timothy Holmes, they state: "None of our general hospitals receive smallpox. Most of them exclude scarlet fever and measles. Several exclude by rule all cases of fever, viz.: The London, University College, Charing Cross, and the Royal Free." They also state that many of the London hospitals (viz. St. Bartholomew's, Guy's, St. Thomas's, Middlesex, and St. George's), and a few country hospitals (Bath, Chichester, Reading, and Oxford), admit all febrile diseases except smallpox, and in some scarlet fever, and that their experience sufficiently showed that such diseases might be treated without excessive danger in the wards of general hospitals. They, however, stated that in the great majority of country hospitals such cases were treated in separate buildings

H

or wards. They moreover admitted that the presence
of infectious disease in a general ward was attended
by a certain amount of risk, that typhus fever showed
some tendency to spread among the patients, and
that at the time of their visit to the Middlesex
Hospital two patients were suffering from scarlet fever
caught in the hospital. It is, however, probable that
the absence of better means for the isolation of infectious disease at the time, affected the opinion of these
eminent physicians, as the following sentence indicates :
" Now, patients with scarlet fever must be treated
somewhere, and we believe they are much less likely
to spread the disease in a general ward than if kept at
home. In a well-ventilated and not overcrowded
hospital the poison, even when typhus fever patients
are accumulated, extends to a very short distance
indeed around each individual case."

Sir William Gull[1] states that he remembered at
Guy's Hospital they took in 75 cases during the
Irish famine, and that they did very well in the
ordinary wards. The disease did not spread. They
were scattered, and by being scattered, they were not
infectious.

Dr. Howse states that cases of typhus in former
epidemics have been placed and treated success-
fully in the large wards at Guy's amongst
the other patients, and that a certain number of
cases of typhoid and diphtheria are still admitted

[1] Smallpox and Fever Hospitals Commission, 1882, Minutes of Evidence,
Questions 4470-73.

into the general medical wards with favourable results.[1]

I have myself, when acting in the Fever and Smallpox Hospitals at Homerton, London, seen a number of cases of typhus treated in the same ward with about an equal number of mixed non-infectious patients, and without any spread of the disease. This, however, is only possible under the most favourable conditions, such as perfect ventilation, abundant floor and cubic space, and careful management and nursing. Even in our best hospitals, one or more of these conditions fail sometime or other, with the result that the disease spreads, as the following facts show. In 1849,[2] 3 patients suffering from typhus fever in the Edinburgh Royal Infirmary gave the disease to 7 of 38 patients in the ward. In St. Bartholmew's, London, in 1838, 11 visitors, 16 nurses, and 21 patients admitted for other affections, suffered from typhus fever. At Guy's in 1862 one or two patients suffering from typhus fever were admitted into a ward of 50 patients, and 7 took the disease. Infectious disease was found to spread in St. Thomas's in 1865, in the Charing Cross in 1862, in the Westminster in 1865, and in the Middlesex in 1867. In the four years 1862-65, 407 cases of typhus fever were admitted into Guy's, St. Thomas's, German, King's, and the London hospitals; 24 patients who had suffered from

[1] Article on "Hospital Hygiene," *Treatise on Hygiene and Public Health* (Stevenson and Murphy), p. 795.
[2] *Continued Fevers*, Dr. Murchison, 3rd ed. pp. 144 and 690.

other diseases and 56 officials contracted the disease, 103 died, giving a mortality of 21.15. 7498 cases were admitted into the London Fever Hospital during the same period; 30 patients who had suffered from other diseases and 81 officials contracted the disease, 1413 died, giving a mortality of 18.57.[1]

In the Pendlebury Hospital, Manchester, in 1877, the total admissions were 560. Of these 183 suffered from infectious diseases; 18 patients in the general wards contracted scarlet fever, 6 patients admitted for measles contracted scarlet fever, and 3 admitted with scarlet fever contracted measles. In 1878, of 901 patients admitted 204 were infectious cases; 14 patients in the ordinary wards contracted scarlet fever and 3 measles, and 8 patients under treatment for measles contracted scarlet fever. In 1879, of 859 patients admitted 136 were infectious cases; 12 patients in the ordinary wards contracted scarlet fever, and 3 contracted measles. In 1880, of 840 cases admitted 131 suffered from infectious diseases; 12 patients in the general wards contracted scarlet fever and 3 measles, while 3 patients who had suffered from measles contracted scarlet fever, and 4 patients who had suffered from scarlet fever caught measles.[2]

In this hospital the infectious cases admitted were suffering from "fever," measles, or scarlet fever, and were treated in one special fever ward. Dr. Thorne Thorne states that "this hospital is constructed on

[1] Murchison, *Continued Fevers*, 3rd ed. pp. 691-692.
[2] Supplement to Tenth Annual Report of the Local Government Board, p. 184.

the most modern principles, and subject to a stringent administration . . . that fever patients in their passage from the fever reception room to the fever ward do not pass within 45 feet of any of the wards occupied by patients . . . that it would be difficult to find any general hospital where the reception of infectious cases is subjected to stricter regulations with a view to their isolation, and that there is probably none where the conditions affecting the patients themselves are more conducive to this end."

One is, therefore, forced to the conclusion that infectious diseases should not be treated in a general ward, or even in a separate ward, in connection with a general hospital, and also that it is necessary to have separate wards for as many infectious diseases as are to be treated at one time in an Isolation Hospital, that these wards should be at safe distances the one from the other, and that the greatest care is necessary both in designing and administering such hospitals to prevent patients suffering from one disease from infecting those suffering from another infectious disease.

Authorities agree as to the necessity, in all infectious hospitals, of having the administrative building detached at a safe distance from the wards; of having ward blocks, intended for different diseases, one storey high, detached, and at a safe distance the one from the other, and of having certain outhouses, such as laundry, stores for infected clothing, disinfecting chamber, mortuary, ashpit, coal-store, etc., erected at a safe distance both from the wards and from the administrative building. The object that should be aimed at is to render the isolation of the sick as complete as possible, not only from the danger of infection from other diseases, but also from all sources of foul air. The patients and attendants should be guarded against air, food, or water being contaminated in the administrative building, as well as against air, food, water, or clothing, or any other material, being contaminated by effluvia or specific infectious agents in the outhouses, or arising from the sick in the same ward, or in any other ward within the hospital grounds. In order to prevent communication between the inmates of an Isolation Hospital and the public, the hospital grounds should be surrounded by a stone wall at least 6 feet 6 inches high.

" The constituents of the atmosphere intermingle chiefly in a horizontal direction, hence the necessity of combating by increased space the effect of contact, and of the proximity which constitutes crowding, and the influence which is communicated from patient to patient, ward to ward, and from one building to another. To strive successfully against contagious influences not only needs an increase of the cubic space allowed to each sick person, but in addition, and above all, an increase of superficial area. . . . For similar reasons the multiplication of stories is contraindicated, as each of them generates a more or less vitiated atmosphere. . . . Buildings completely isolated, and all having the same aspect freely exposed to light, and to the action of the wind and rain, should be placed on a single, or in parallel lines, at intervals of 101 to 119 yards, so as to secure effective separation and external aeration."[1]

Dr. Wylie,[2] of New York, states that "experience and science agree in showing that widely-detached one-storey wards allow the most thorough ventilation, and therefore the smallest chance for the accumulation of infectious particles.   They make classification of cases easy and natural.   They require less vigilance, dust and foul air find fewer lurking holes and channels, cleanliness and ease of supervision, as well as fresh air, are more readily secured."

Miss Florence Nightingale states: "The most healthy hospitals have been those on one floor only, and this because they require less scientific knowledge and practical care in ventilation. . . . There are instances in which the mortality of patients on upper floors has been higher than of those on the floors below. . . . If another floor is added a community of ventilation exists between the ward below and the ward above by the common staircase, and by filtration of air upwards through the floor."[3]

" The mode of construction in hospitals is, it is presumed, to be determined by that which is best for the

[1] Extract from Conclusions of Surgical Society of Paris, Mouatt and Snell's *Hospital Construction and Management*, pp. 21 and 22.

[2] " Hospitals, their History, Organisation, and Construction," see Buck's *Hygiene and Public Health*, p. 753, vol. i.

[3] *Notes on Hospitals*, Miss Florence Nightingale, p. 58.

recovery of the sick. If any other consideration is
taken, such or such a percentage of mortality is to be
sacrificed to that other consideration, but it so happens
that the safest for the sick is in reality the most economical mode of construction."[1] . . . "The first principle of hospital construction is to divide the sick
among separate pavilions. By a hospital pavilion is
meant a detached block of building capable of containing the largest number of beds that can be placed
safely in it, together with suitable nurses' rooms, ward
sculleries, lavatories, baths, water-closets, all complete,
proportioned to the number of the sick, and quite uncon-
nected with any other pavilion of which the hospital
may consist, or with the general administrative offices,
except by light airy passages or corridors. A pavilion
is indeed a separate detached hospital which has, or
ought to have, as little connection in its ventilation
with any other part of the hospital as if it were really
a separate establishment miles away."

Sir Douglas Galton states that "it may be taken
as accepted as a rule that, so far as the sick are concerned, they would be better placed in wards all on
one floor opening out of a common corridor, and if the
land is cheap, and the site fairly level, it is probable
that such an arrangement might be more economical
than building two-storied buildings."

Dr. Caspar Morris[2] and Messrs. Mouatt and
Snell quote the opinions of the Chirurgical Society

[1] *Notes on Hospitals*, Miss Florence Nightingale, p. 59.
[2] *Hospital Construction and Management* (John Hopkins), p. 181.

of Paris that not less than 64 yards per patient
should be allowed (non-infectious hospitals evidently
considered). Dr. Stephen Smith[1] remarks that
"where grounds are ample . . . there is no reason
why the number should exceed forty persons per acre,
as a general average. . . ."

"An ideal site for an hospital would be where the
conditions of soil, subsoil, drainage, water supply, and
all surroundings were most free from local causes of
impurities, and where there were fewest buildings and
habitations to exclude or intercept air and light, or
to be themselves active agents in the creation of
causes of unhealthiness, such as factories, workshops,"
etc.[2]

According to Dr. Francis H. Brown,[3] "a hospital
should be surrounded by a zone of aeration unen-
cumbered with buildings, etc., to a distance of twice
its height. "That building makes the best hospital
which is situated high, dry, and detached, with sufficient doors and windows for cross ventilation, open
fireplaces, secure roofs and walls, easy access, lofty,
and of moderate size."

Dr. Thorne Thorne[4] states : "I have been led to
regard an elevated site on a gentle slope, and on a dry
soil where the free circulation of air about the hospital
buildings was not interfered with, and where an

---

[1] *Hospital Construction and Management* (John Hopkins), p. 284.
[2] *Hospital Construction and Management*, p. 22, Messrs. Mouatt and Snell.
[3] Buck's *Hygiene and Public Health*, vol. i. p. 738.
[4] Report of Medical Officer, Local Government Board Supplement, 1880-
1881, pp. 8 and 11.

abundant and wholesome water supply, with reasonable facilities for drainage, were available, as having distinct advantages over sites differently circumstanced. . . . Amongst the hospitals visited, and where these several necessary conditions have been fulfilled, I would name those in Bradford, Cheltenham, Tunbridge, and Warrington Urban Districts, in the Berkhampstead and Solihull rural districts, and that belonging to the Weymouth Port Sanitary Authority. In none of these instances does the number of patients per acre exceed twenty."

" It is, as a rule, undesirable to select any site which is less than some two acres in extent, and even then regard should be had to the need for extension of hospital buildings, whether for temporary purposes, or owing to increase of population." [1]

Some appear to be of opinion that infectious diseases may be prevented from spreading by means of temporary, movable huts, and in some districts in rural Scotland these have already been provided. Corrugated iron and wooden buildings, constructed in sections and ingeniously made for packing, are supplied by various firms for this purpose. These buildings can be erected, taken down after use, packed and transferred to another locality for re-erection.

The advantages claimed for such buildings are that the initial expense is less, while the district will

[1] Paper read in the Architectural Section of the Seventh International Congress of Hygiene and Demography held in London, August 1891 (Dr. Thorne Thorne).

be equally well served. Instead of patients being conveyed long distances to the hospital, the hospital is to be brought to them.

The scheme at first sight looks well both from a financial and medical point of view. Against it, however, must be set the expense of demolition, packing, conveyance, and re-erection of such a building every time it is moved, the additional wear and tear of the structure, the difficulty of providing a building which will give sufficient shelter, and afford the necessary comfort and safety for isolation, and for the recovery of the sick in our climate, as well as the delay which will necessarily take place in selecting a fresh site in every locality where it is proposed to have it temporarily erected. An ambulance carriage will be required whether the hospital is movable or stationary, as it cannot always be erected within a short distance of all the infected families.

The primary use of an Isolation Hospital of any kind, whether movable or fixed, is to prevent the spread of disease by providing means for the immediate seclusion of such cases as may occur. The delay which will necessarily take place in getting a movable hut ready for the reception of patients will allow the disease time to spread.

Even in thinly populated localities permanent structures should be provided, and be always kept in readiness for the admission of patients. In remote localities where outbreaks of disease are rare, and where two infectious diseases but seldom break out at

one and the same time, it will suffice, at any rate to
begin with, to make provision for the isolation of one
disease in the hospital.   The current expense will also
be less, owing to the fact that nurses need not be
permanently employed in such hospitals.

In addition to pollution by the infectious agents of
disease, an Isolation Hospital is liable to contamination by the effluvia from the bodies, the breath, and
excretions of the patients, as well as from foul linen,
bedclothes discharges from sores, etc. etc.   It is therefore essential to construct and furnish a hospital of
such material as will not absorb noxious effluvia, and
to provide sufficient floor and cubic space, a plentiful
supply of water, perfect ventilation, proper arrangements for the removal of excretions and dirt in its
various forms, and for disinfection and washing.
Provision should also be made for the safe isolation
of nurses and servants attending to each disease, and
for their healthy accommodation whether on or off
duty.   There must also be proper and safe means for
the cooking and distribution of food, as well as for the
storage of food, clothes, medicines, and other necessaries.
At the same time the whole building should be designed
in such a way as to render the efficient discharge of
their duties as convenient as possible for all engaged
in the hospital.

An Isolation Hospital of whatever size should
consist of—(1) A detached administrative block or building, containing the kitchen with its necessary stores,
etc., and the living and sleeping apartments of the

hospital staff; (2) Wards, with their appendages in separate pavilions, or blocks, or cottages, at safe distances, providing for the separation of the sexes, and for patients suffering from different diseases; (3) Outhouses such as laundry, stores, disinfecting apparatus, mortuary, etc.

The administrative block should minister to the whole hospital (excepting perhaps when smallpox[1] is isolated). It should at least be forty feet from the wards. No food, clothing, earthenware, furniture, or anything else should get mixed in the work of administration. The block or ward used for one disease should be at least forty feet from the ward occupied by patients suffering from any other disease.[2] When a patient is admitted,[3] he should be sent into the receiving room of the block set apart for his disease. If it is a matter of doubt whether the patient is suffering from an infectious disease, or if infectious, what disease, he should be sent to an isolation ward until the nature of the disease becomes more apparent. If there is no doubt about the case, it should be sent to the ward intended for it. The patient, if able to bear it, should be undressed in the bath-room, bathed, provided with clean night-dress, and put to bed. The

[1] The nurses attending smallpox should be accommodated in rooms apart from the general administrative building. If sleeping in the administrative building, they should change their clothes and take a bath on going off duty.

[2] Such diseases as typhoid, diphtheria, and typhus fever might, with care on the part of the attendants, be treated without danger in wards separated by a shorter distance than forty feet.

[3] Every case of infectious disease removed to a hospital should be certified by a medical attendant, and the consent of the friends or relatives obtained if possible.

clothes should immediately be put into the disinfecting chamber, and after being disinfected and washed stored in a cupboard or apartment provided for it.   If necessary the clothes should be destroyed.   The ambulance should be at once disinfected.

All bedclothes, linen, towels, etc., should be marked separately for each disease, stored in its special department, and on no account used in any other part of the building.   Care should be taken to prevent the clothes from getting infected in the laundry.   The clothes from each department should be thoroughly disinfected before being taken to the laundry, and the laundry so constructed that it cannot retain infection and become a source of danger to the hospital.   When dry the clothes should at once be brought back and stored in their special department.

All earthenware, spoons, knives, forks, feeding glasses, and other articles, should be of different patterns for each disease, or differently marked for each department, and washed and stored there.   The main articles of food should be conveyed to each block in utensils belonging to the administrative department by a person not engaged in any of the wards, and there transferred into utensils belonging to that block.

The nurse should keep to one disease exclusively when on duty.   She should wear a loose wrapper and leave it in the apartment provided for it when going away.   She should wash her hands well with some efficient disinfectant, and if necessary take a bath. Her bedclothes, linen, etc., should be kept in her own bedroom and not mixed with those of a nurse attending any other disease. When a patient is to be discharged and there is not a special discharging room, the bath and bath-room should be rendered as free as possible from infection, and it should be so placed that after being bathed and dressed the patient need not return again to the infected ward. In the hospital ships, *Atlas* and *Endymion*, in 1881 a patient before being discharged was taken from the *Atlas*, on which he was treated, by a drawbridge to the *Endymion*. He entered a bath-room, and after taking his bath he went by a door into a passage leading into the dressing-room. In this passage a large carbolic spray was playing. In the dressing-room he found a new suit provided for him. After dressing he was sent home direct from the *Endymion* by means of conveyances provided by the Metropolitan Asylums Board.

In the Blegdam Hospital, Copenhagen, "two rooms are provided for visits to convalescent patients. The wall of separation is furnished with a window and a grille. The patient is in one room and his visitor in the other, and they can see and converse but cannot touch one another. The access to the patient's room is quite separate and distinct from that to the visitor's room.[1]

In the Belvidere Hospital, Glasgow, undressing

---

[1] Burdett, *Hospitals and Asylums of the World*, vol. iv. pp. 283 and 284.

closets, bath-rooms, and a dressing-room are provided at a distance from the wards, for patients to be discharged.

Often but little attention is paid to all these considerations in designing an infectious hospital, but if not attended to, disease is liable to spread not only among the inmates, but from them to the public outside. It is, however, practically impossible to provide all these conveniences in small hospitals in rural districts.

Such hospitals, however, are in many ways different from large institutions in the towns. In rural districts the hospital is often empty. It is very rarely full. Although provision may be made in it for more than one disease, it is very rarely that many patients suffering from even two diseases are isolated in it at the same time.

I shall be able to show further on that a room or ward used for the isolation of one disease may, by careful disinfection, be afterwards rendered perfectly safe for the treatment of another disease. If, therefore, a small hospital, instead of two large wards, consists of a number of small rooms sufficiently separated, more than one disease may be isolated in it at one and the same time, and a room can always be kept for doubtful cases. An empty room may be used as a discharging room, or, if the bath-room is placed in a suitable position, it can be sufficiently disinfected to serve as a discharging room when necessary. " The larger the wards the more difficult it is to secure the isolation and classification of patients, and the smaller the wards the greater is the labour and cost of supervision and attendance."[1]

---

[1] *Hospital Construction and Management*, Dr. Billing (John Hopkins), p. 40. With small wards less expense is incurred in firing and lighting, when only one or two patients are isolated.

CHAPTER VI

Hospital Construction

THE administrative block should in size and arrangement be in proportion to the size of the hospital, the
number of patients and others to be served by it, and
the staff to be accommodated in it.   Authorities
appear to be unanimously of opinion that whatever
may be the size or the material of which the wards are
built, the administrative building should be a permanent, well-built structure.   In growing towns and
rural districts it should as a rule provide accommodation and conveniences in excess of the immediate
requirements.   This is not only desirable, in order to
save expense should the hospital at any time be en-
larged, owing to increase of the population, but also to
provide such extra accommodation as may be found
necessary for nurses during any sudden emergency
from an exceptional outbreak of disease.   The administrative building should be erected within the hospital
grounds at a safe distance from the wards and outhouses (at least forty feet), and it should be so placed
as to be convenient for the wards, yet so situated that
persons, such as visitors and friends of patients, may

enter it from the front gate without coming within reach of infection from the wards or any other part of the building. In our climate it is generally desirable to have it connected with the wards by a roofed passage or corridor, open as much as possible on both sides.

The kitchen, with its accessories, such as stores, larder, pantry, etc., should be conveniently placed for the wards, and in size and arrangement should be proportionate to the requirements of the hospital as a whole. A room should be provided for the medical officer, in which, if necessary, medicines and other appliances may be kept. There should also be a store for linen used in the administrative building, in which new clothes and other articles might be kept until required in the wards. Such articles, however, when once used should not be returned to this store, but kept in places provided in the special block or department to which they have been sent. In most other respects the administrative cottage should be built and arranged according to the rules and regulations in regard to water supply, drainage, ventilation, cubic space, and other conveniences, which govern the construction of good, modern dwelling-houses of the same size.

Wards.—Of all parts of a hospital building the wards are the most important. Their structure and arrangement, therefore, deserve the greatest care and consideration. The reasons for this have already been given. This is necessary even in the case of wards

intended for non-infectious diseases, but it is much more
so in Isolation Hospitals, where, in addition to the
various deleterious emanations which are generated in
persons suffering from non-infectious ailments, specific
agencies are liable to accumulate and to contaminate the
air as well as the building and furniture. The whole
hospital should be guarded against the absorption and
retention of these emanations and agents of disease,
but more particularly the wards, for these are exposed
to unhealthy influences in a more concentrated state,
and more continuously than the other parts of the
building. I have already shown that various impurities
have been found floating about in the air of hospital
wards and on and within the walls ; that bacteriologists
have conclusively proved that the organisms of
disease are able to live outside the human body for
an indefinite period, and may retain their power of
infection ; and that other observers in the field of
preventive medicine have proved the occurrence of
outbreaks of infectious diseases as the result of infection remaining in and about dwellings for varying
periods.

In small hospitals such as those under consideration, each ward block, pavilion, or cottage, should at
least contain two wards, one for males and one for
females, and a nurse's duty-room, as well as proper
provision for cleanliness, bathing, and for the disposal
of excretions. In some cases it is desirable to have
more than two wards in one block in order to allow of
the separation of a doubtful, a troublesome, or a dying

case, from the others, or for the accommodation of a paying patient.

The ground on which a hospital pavilion is built should be covered with concrete or asphalt, and the foundation built in arches so as to allow the floors of the wards to be elevated from the base by six to eight feet.[1]

This has been done in the hospital at Ealing[2] and other places. It prevents the possibility of the soil immediately under the wards from becoming infected by the accumulation of filth, or by soakage from drains, or in any other way. When this is not done the ground should be drained and covered with a thick layer of concrete. A damp-proof course should be laid on the foundation, and the space under the floor should be properly ventilated.

The position of the nurses' duty-room and the other ward accessories, such as bath and water-closets, will be seen to vary slightly in different well-constructed hospitals, by reference to the plans appended.

The materials of which ward pavilions are built also vary. Brick, stone, concrete, wood, corrugated iron, are used in different localities. Most authorities appear to be agreed as to the superiority of permanent brick or stone buildings for localities that can afford to bear the expenditure. It is doubtful, considering the wear and tear and the extra expense of the upkeep of buildings constructed of less enduring material, whether

---

[1] Francis H. Brown, M.D., Buck's *Hygiene*, vol. ii. p. 747.
[2] See Plan, Appendix.

stone or brick would not be cheaper for all localities, more particularly as the payment may be extended over a long series of years. Some authorities, however, appear to be in favour of wooden buildings. Dr. Francis Brown of New York states that the plan which commended itself most favourably to medical men of late years in establishing a hospital, whether of large or small dimensions, has been to build it of detached wooden pavilions, with an administrative building of more permanent material, and that it cannot be questioned that wooden buildings "bring the patient nearer the condition of nature," that such an arrangement is more satisfactory on the ground of expense, and that a barrack hospital is more speedily and easily erected.[1] He quotes Dr. Billings and Sir Douglas Galton as being in favour of his views. It may, however, be observed that since Dr. Billings wrote the article quoted by Dr Brown, he has modified his view on this question. In the able report written by him for the Trustees of John Hopkin's Hospital, Dr. Billings states: "But the statement that this temporary character should be adopted for all hospitals, and especially all parts of hospitals, was, I am now satisfied, too sweeping. That is to say, I do not think it necessary that all the buildings of a hospital should be destroyed or removed at certain regular intervals, in order to prevent infection, and there are some things to be taken into account in favour of more permanent structures under certain circumstances, to which I did

---

[1] Buck's *Hygiene and Public Health*, vol. ii. p. 745.

not give sufficient consideration." He, however, continues : "Barrack hospitals are best suited for Government purposes, for contagious and infectious diseases, and in general where much subdivision of patients is unnecessary, where number of attendants and cost of fuel is not taken into account, and, as I have above remarked, for hospitals which are liable to be mismanaged."[1] Here, again, Dr. Billings had clearly in view the use of wooden or barrack hospitals for an epidemic of one disease breaking out among soldiers, as it is only under such conditions that much subdivision of patients would not be necessary. As already stated, the requirements of Isolation Hospitals for small towns and rural districts, such as those under consideration, are about the very opposite to what Dr. Billings had in view. "Subdivision of patients" is absolutely necessary for the separation of the sexes as well as for the different diseases. The "number of attendants and cost of fuel" forms a very important item in the expense, and a serious consideration in the management of such hospitals, and instead of being "liable to be mismanaged," in what I conceive to be Dr. Billings' use of the term, no other kind of hospital requires the same assiduous attention, for not only are the patients admitted with a view to receive the best treatment and to be put under the most favourable conditions for recovery, but the greatest care is necessary to prevent the diseases from spreading from patient to patient and from them to the attendants and others.

[1] *Hospital Construction and Management* (John Hopkins), pp. 16 and 17.

The quotation from Sir Douglas Galton[1] is as follows, " I would add one more caution. Do not build for a long futurity. Buildings used for the reception of the sick become permeated with organic impurities, and it is a real sanitary advantage, that they should be pulled down and entirely built on a fresh site periodically."

Surgeon-General Billings and Dr. Brown appear to think it necessary that such buildings should not be used for longer than fifteen years, and Dr. Billings states that he had seen a decided tendency to erysipelas and pyæmia in a characteristic barrack ward, within six weeks after it was first occupied.

The history of the hospital ship *Dreadnought* and its successor is of some interest as bearing on this question. In 1858, Sir John Simon states, " The old hospital ship *Dreadnought* has acquired a very evil reputation for the prevalence of these infections, dependent, no doubt in part, on the natural ill-adaptedness of a ship to the purposes of a hospital, but probably also, in part, dependent on organic contamination still lingering in the wooden walls of the wards. Early last year another commodious ship was substituted for the *Dreadnought*, and Mr. Tudor, the resident surgeon, informs me that whereas in the two years preceding that change nine out of twenty-two amputations had terminated fatally, only one amputation had proved fatal out of sixteen

[1] *Construction of Hospitals*, Sir Douglas Galton, being an address delivered before the British Medical Association at Leeds, 1869.

performed in the year following the change, and that whereas formerly erysipelas and hospital gangrene were so common and so spreading as to have let him see there at one time as many as eighteen cases of hospital gangrene, he has now scarcely seen erysipelas except in patients admitted with it, from whom as a rule it no longer spreads to the other inmates of the wards. . . . Little by little since that time the old state of things has returned, and now the report made by the ship is, that traumatic infections are 'excessively frequent,' that 'operations do badly,' that pyæmia is frequent."[1] It thus appears that within a few years this hospital ship had become infected with the micro-organisms of various diseases to such an extent as to infect the inmates, just as the decided tendency to erysipelas and pyæmia appeared in Dr. Billings' barrack wards within six weeks.

This would lead one to believe that the destruction of such buildings would become necessary in a much shorter time than fifteen years, and if this was done as often as necessary, it is doubtful if these buildings could be recommended on the "score of expense." If instead of surgical cases being treated in such hospitals, cases suffering from one infectious disease were admitted at one time, and of another infectious disease at another time, the probability is, that the walls would soon become so infected by the organisms of the various diseases as to defy disinfection, and that patients isolated for one disease would frequently fall

[1] Report of Medical Officer to Local Government Board, 1863, p. 64.

victims to other diseases contracted in the wards.   If special wards were always used for each special disease this would not occur, although the effect of other evil influences might be felt.

To provide special wards[1] for every infectious disease in thinly populated rural districts or small towns would be impracticable, as I have already endeavoured to show.   If, however, the inner lining of a wooden hospital ward consisted of some impermeable material which could in no way harbour organic or infectious germs, as recommended by Dr. Brown,[2] such buildings might be made safe for the reception of cases of different infectious diseases. Under such circumstances, I do not see the advantage of a wooden outside wall, over stone or brick.   In our climate more particularly such would require careful and frequent painting, and the expense of up-keep, if added to the interest of the primary outlay, would probably in time come to equal that of the more expensive stone or brick building.

Corrugated iron has been recommended by various firms, as being more durable than wood, and some hospitals have been erected, both in England and Scotland, of this material.   As hitherto constructed in England, neither corrugated iron nor wooden hospitals appear to have given satisfaction.   Dr. Thorne

[1] The wards of the N.W. Hospital of the Metropolitan Asylums Board consist entirely of wood, or corrugated iron and wood.   Scarlet fever, typhoid fever, and diphtheria are treated there ; 1692 cases in 1892.   Dr. Gayton, the Medical Superintendent, however, informs me that the same wards are always used for the same diseases.

[2] Buck's *Hygiene*, vol. ii. p. 747.

Thorne states, after a careful examination of such hospitals, "that, having reference, however, to the experience that has been acquired, I cannot but conclude that, as regards permanent hospitals in this climate, wooden or iron buildings as ordinarly constructed are not as a rule well adapted to the purposes of wards. That they can be constructed so as to ensure a reasonable and a fairly equable ward temperature, I do not doubt; but when so constructed, their original cost would probably not fall short of, if it did not exceed, that incurred in the erection of ordinary brick buildings; they would be less durable than these, and the cost of maintaining them in a proper state of repair is undoubtedly greater than that needed for the maintenance of the more substantial structure."[1]

A firm in Scotland has recently patented an improved system of constructing weatherboard and iron hospitals. Two layers of highly non-conducting material are placed in the wall, "converting the space between the exterior and interior coverings into two air compartments, which air compartments are insulators of themselves." It is stated that in this manner an equable temperature can be easily maintained in such buildings at all seasons. The Scottish Board of Supervision sanctioned loans to local authorities for the erection of iron or weatherboard hospitals on this patent. Portable hospitals of corrugated iron are also constructed

[1] Supplement to Report of Medical Officer to Local Government Board, 1880 to 1881, p. 10.

by the same firm. I am not myself satisfied that an
equable ward temperature could be maintained in these
temporary movable buildings. I saw one of the permanent hospitals, but before patients were admitted to
it. It looked very well, though I felt doubtful as to
the advisability of providing an inner lining of wood,
V grooved, at the junction of the boards. The object
of the V groove is to conceal the gaping which is
caused by contraction of the wood, a very undesirable
thing in an infectious ward. If such hospitals are to
be internally lined with wood, a plain surface would be
preferable. The least crack would be at once noticed.
The monotony of the wall might be relieved by means
of dadoes and panelling. The fissures between the
deals should be filled with cement, putty, or strips of
wood to prevent infected dust from being carried by
currents of air into the wall. There are various ways
in which a wooden floor or wall might be dressed and
finished. If thoroughly rubbed with glass or sand-
paper, and brushed over with shellac dissolved in
spirits of wine, and this repeated several times, the
wood would get hardened, and might in the case of
floors be kept easily polished by being rubbed occasionally with a mixture of bees-wax and turpentine
(one pint to a quarter pound[1]), or it might be made
impermeable, antiseptic, and easily washed by the
application of coal tar, mixed with heavy coal oil, in
the proportion of one-fourth of coal oil to three-fourths
of coal tar, by weight, not by measure. Before apply-

[1] Dr. Howse's *Treatise on Hygiene and Public Health*, p. 781.

ing this mixture, the floor or wall should be well
scrubbed with a hard fibre or metal brush, then washed
with soda, and finally impregnated with a solution of
one in one thousand of corrosive sublimate. When dry,
a first coating of the coal tar mixed with coal oil should
be applied by means of a painter's brush, and spread
as thin as possible with a hard brush. In forty-eight
hours another coat should be applied, and in another
forty-eight hours a third and last coat. To keep the
floor in good order, it is enough to sweep it and to
wipe it now and again with a floor cloth, wrung out in
water or antiseptic solution, and when dry to rub it up
with a hair or wool brush, on which some drops of
petroleum have been sprinkled. The smell of the coal
tar disappears in about three days.[1] /

If stone or brick is used for the wall of a ward
block, the wall should be hollow, so as to prevent
dampness. The inner face of the wall should be
covered with some material which would present an
impervious polished surface, incapable of absorbing
organic matter, or of affording shelter to the micro-
organisms of disease. Wall plaster, wood, paint, var-
nish, and wall paper absorb organic impurities, as I
have already shown.

Sir Douglas Galton[2] states that enamelling the
walls like panels of carriages would probably make
them impervious, but that it is rather expensive and
liable to be scratched and damaged. Most authorities

[1] *Sanitary Record*, November 15, 1892.
[2] *Construction of Hospitals*, an address delivered before the British Medical
Association at Leeds in 1869.

appear to be in favour of glazed brick or Parian cement. These require to be applied direct to stone or brick walls, without any intermediate lathing; hence the necessity of building the wall hollow. Parian cement is rather costly, and its want of elasticity is unfavourable to its use in ceilings.[1] It is liable to crack, and the colour is not uniform.[2]

Prof. Jones[3] states that coloured tiles or large porcelain or glazed earthenware slabs, joined perfectly by means of good cement, might be used. A writer in the *Philadelphia Medical Times*, quoted by Dr. Brown in his article on " Hospital Construction and Management " in Buck's *Hygiene*, states that glass would seem to meet the requirements, being incapable of absorbing organic matter, not expensive, strong when sufficiently thick, impervious to water or dampness, and could be made of suitable colours.[4] Miss Florence Nightingale thinks that the best material for woodwork in wards is polished or varnished wainscot oak. This would also probably be considered too expensive. Mr. Percival Gordon Smith[5] states that glazed bricks have been used with good effect if set with fine joints in white lead ; and Dr. Thorne Thorne[6] is of opinion that the internal surface of walls is best made of glazed bricks

[1] Sir Douglas Galton.   [2] Miss Florence Nightingale.

[3] *Hospital Construction and Management* (John Hopkins), p. 112.

[4] Dr. Snellen of Utrecht informs me that the inside of the walls of the operating room in the Rotterdam Hospital is covered from floor to ceiling with large plates of glass.

[5] "On the Planning and Construction of Hospitals for Infectious Diseases," *Epidemiological Society Transactions*, vol. ii. 1882-83, p. 142.

[6] Supplement to Report of Medical Officer, Local Government Board, 1880-81, pp. 12, 13.

or cement; that the glazed brick walls, as in the Delancey Hospital, Cheltenham, are attractive and admit of easy cleaning, and that the Parian cement, if put on so as to present a hard, impervious, porcelain-like surface, as at Weymouth, is well adapted to the purposes of hospital wards. In the absence of Parian cement or glazed bricks, the safest arrangement is probably ordinary brick, cement, or lime plaster, period-ically scraped so as to remove the tainted surface and be again lime-whited or painted.[1] If the walls are plastered with a view to being painted, care should be taken that the painting is not done before the plaster is perfectly dry.

To secure an equable temperature in our climate, the walls should be at least fourteen inches thick. The angles at the junction of the ceiling with the walls, of the walls with the floor, and of the end with the side walls, should be rounded concave, so as not to harbour dust and to be easily cleaned. Cornices and other projections should be avoided.

So far as I know, the shape universally adopted in this country for Isolation Hospital wards has been rectangular. Prof. Marshall, who first advocated the advantages of circular wards, gave a description of the Antwerp Hospital in the *British Medical Journal* for August 1882. "A circular ward would be uni-formly free to all quarters of the compass except at one point, where an open or partially open corridor would connect it with the other wards or with the

---

[1] Sir Douglas Galton on *Construction of Hospitals.*

offices, so that this form must give free frontage, and must also permit of the freest possible access of light and air. If it be true that 'the worst wards are those where least light and air are provided,' and that 'a closed court, with wards around, is the worst arrange-ment,' then, since a circular ward is the inverse of this last-named plan, it would follow that this form is best adapted to fulfil the desired conditions ; its uniformly rounded exterior, receding from all adjacent buildings, would receive light, air, and wind from every direc-tion. There is, moreover, the least possible interfer-ence with light and air to other buildings. The circular form also offers certain advantages with regard to wall space, floor space, and cubic space for each patient. It is not adapted for wards with less than eight beds."

With regard to the cost of construction, Mr. Percival Gordon Smith states, that while the walls would cost rather less to erect than in the case of rectangular wards, the flooring and roof would be more costly. The window sashes, glass, and doors would not need to be curved, and, on the whole, he believes that the difference of cost between a circular block of wards and an ordinary pavilion for a corresponding number of beds, with equal space per bed, would be but slightly in favour of the straight building. As to the artistic possibilities of a hospital with wards on the circular principle, Mr. Smith expresses himself very decidedly. " In skilled hands," he says, " it would lend itself in the most happy manner to the production of buildings which would, undoubtedly, be the pride of the towns possessing them." The freest exposure to light and air, the largest amount of floor, wall, and cubic space, are even more necessary for infectious than general diseases. Circular wards, therefore, appear to answer the requirements of an infectious hospital, as well as rectangular wards of moderate size. Appended is a plan of a circular ward for the isolation of small-pox, as suggested by Dr. Burdon Sanderson to the Infectious Hospital Commission.

The difference in the cost of circular and rectangular wards, according to Mr. Percival Gordon Smith[1] may be safely said not to exceed about 2 per cent.

Dr. Louis Parkes draws attention to the fact that if a circular ward is to accommodate the same number of patients as an oblong ward, having an equal floor measurement and cubic contents, the beds of the patients, which are placed around the wall, must be very closely packed together, and that 8 feet of wall space per bed cannot, by any possibility, be attained.[2]

In large hospitals, whether for infectious or non-infectious diseases, the size of the wards is governed, to a large extent, by the number of patients that can be attended to by one nurse. In small hospitals, however, the size of the wards should depend more on the classification of patients, according to sex and disease, and in some hospitals including arrangements for the reception of private or paying patients.

[1] Article on "Dwellings" in *Treatise on Hygiene and Public Health* (Stevenson and Murphy), vol. i, p. 729.

[2] *Hygiene and Public Health*, Louis Parkes, p. 429.

K

All authorities appear to be agreed as to the necessity in isolation wards of allowing 2000 cubic feet of air space per patient; 144 square feet of floor space [1] and 12 feet of wall space to each bed. Mr. Percival Gordon Smith states that it might be better, if intended for a row of beds on each side, to have the wards 26 feet wide, and 13 feet high, than to be 24 feet wide and 14 feet high. [2]

The best material for flooring [3] is hard wood, such as oak, hard pine, red deal, or ash well seasoned. Cement or tiles are not suited for cold climates. All joints in the floor, as well as in the other woodwork in the wards, should be tongued and grooved, and put together with white lead. It should be saturated with oil and resin, or treated as previously described. Miss Florence Nightingale [4] says that well seasoned oak is the best, treated with bees-wax and turpentine. Francis Brown [5] states that the floor may be oiled, or treated with paraffin, melted and poured on it, and then ironed in with hot irons. Paraffin dissolved in turpentine may also be applied as a paint to the walls and furniture. [6]

---

[1] Dr. Thorne Thorne states that the beds should be so arranged as to ensure for each patient a floor space of 156 square feet. Paper read in Architectural Section, Seventh International Congress of Hygiene and Demography, 1891.

[2] *Epidemiological Society Transactions*, vol. ii. 1882-83, p. 142.

[3] See *Hospital Construction and Management* (John Hopkins); Norton Folsom, p. 67; Billings, p. 29; Caspar Morris, p. 200; Stephen Smith, p. 301.

[4] *Notes on Hospitals*, Florence Nightingale, pp. 69-71.

[5] Buck's *Hygiene*, vol. ii., Francis Brown, p. 748.

[6] "Planning and Construction of Hospitals," *Epidemiological Society Transactions*, P. G. Smith, 1882, vol. ii. p. 142. Galton, *Construction of Hospitals*.

Opposite windows should, if possible, be provided
on both sides of the ward. They should extend from
2 feet 6 inches or 3 feet of the floor, to within 1 foot
of the ceiling. It is desirable, if possible, to have
a window between each bed. Sir Douglas Galton [1]
is of opinion that "one superficial foot of window
space to from 50-55 cubic feet of space [in the ward],
will afford a light and cheerful room ; but that this
depends much on the situation, and upon the walls
being very light colours." One square foot of window
opening to about 70 cubic feet is more advantageous
according to Dr. Thorne Thorne. Too much window
space makes it difficult to regulate the temperature
of wards during the heat of summer or cold of winter.
Dr. Angus Smith, F.R.S.,[2] made careful experiments
regarding this question in the Children's Hospital at
Pendlebury, Manchester. In this hospital the window
space is 1 square foot for every 35 cubic feet space.
Dr. Smith found that, with such a large window space,
it was not possible to keep the ward air pure, and at
the same time equably warm.

All windows should be double sashed, and made
to open from the top and bottom. It might be
desirable to use plate glass in the windows, so as to
economise heat.[3] Every window in a hospital ward
should, in addition to being double sashed, have a part

[1] *Construction of Hospitals*, p. 27.

[2] *Public Health*, Wynter Blythe, p. 559.

[3] Sir Douglas Galton, *Construction of Hospitals.* Dr. Russell, Medical
Officer of Health for the City of Glasgow, informs me that the windows of the
Belvidere Hospital are double glazed, which is found necessary and sufficient
for the maintenance of heat.

above the upper sash, hinged below, provided with
side flaps and made to open inwards. This, while
providing ventilation, prevents down draught.

Ventilation and warming of the ward may well
be considered together, as, in this country, the
most powerful aids to ventilation are open fireplaces
and ventilating stoves, by means of which both
purposes are carried out. In a variable climate, such
as that prevailing in this country, with a high summer
and low winter temperature, often accompanied respec-
tively by calm weather or high winds, the heating
and ventilation of wards, so as to secure the desired
temperature and purity of air, are not so easily
attained as might at first sight appear. The opinion
of Miss Florence Nightingale, a lady of uncommon
experience in hospital management, is of value as
bearing on this question. Miss Nightingale writes
that "Nature affords air to both sick and healthy, of
varying temperature at different hours of the day
and night and season ; always apportioning the
quantity of moisture to the temperature, and providing continuous and free movement (of the air)
everywhere. We all know how necessary the variations of weather, temperature, and season are for
maintaining health in healthy people. Have we any
right to assume that the natural law is different in
sickness ? In looking solely at combined warming
and ventilation, to secure to the sick a certain amount
of air at 60° Fahr., paid for by contract, are we acting
in accordance with physiological law ? Is it a likely

way to enable the constitution to rally under serious disease, to under-cook all the patients day and night, during all the time they are in hospital, at one fixed temperature? I believe not! On the contrary, I am strongly of opinion,—I would go further and say —I am certain that the atmospheric hygiene of the sick-room ought not to be very different from the atmospheric hygiene of a healthy house."[1] . . . "In the wooden hospitals before Sebastopol, with their pervious walls and open-ridge ventilation, in which the patients sometime said that they would get less snow if they were outside, such a thing as catching cold was never heard of. The patients were well covered with blankets, and were all the better of the cold."[1] "Patients in bed are not peculiarly liable to catch cold, and in England, where fuel is cheap, somebody is indeed to blame if the ward cannot be kept warm enough, and if the patients cannot have bedclothing enough, for as much fresh air to be admitted from without as suffices to keep the ward fresh. No artificial ventilation will do this. Except in a few cases well-known to physicians, the danger of admitting fresh air directly is very much exaggerated."

Agreeing in the main with all this, it must be understood that every care should be taken to provide means for heating and ventilating wards, so as to have the temperature more or less under control. Cases are now and then met with in every hospital

---

[1] *Notes on Hospitals*, pp. 75 and 76.　　　　[2] *Ibid.* p. 15.

ward which require special treatment, and it may be desirable, for the sake of such patients, to keep the temperature higher for a time than is necessary for the other inmates.[1] Unless provision is made for efficient ventilation and warming under such circumstances, the benefit that would accrue from the higher temperature might be lost by the increased impurity of the air. Still, to attain this it does not appear to be necessary to provide any elaborate means of artificial ventilation or heating for small hospitals in this country. The condition to be secured is such temperature of the air as may be required at different times, to suit the conditions of the patients under treatment. Slight variations in the temperature are no doubt rather beneficial than otherwise, and a cooler temperature at night, more particularly in the tropics, acts as a tonic. Any one who has visited tropical countries cannot but be impressed with this fact. Yet the conditions in the tropics are different. There, at night, there is a change from extreme heat during the day to a refreshing moderate coolness at night, but in this climate the change may be from a moderate heat or cold during the day to an extremely low temperature at night. In regard to the purity

---

[1] Prof. Jones of the University of Louisiana, and formerly surgeon of the Confederate forces, states, " In convalescents with fever and other acute diseases, the forces having been reduced by the active chemical changes and rapid metamorphosis of the blood and tissues and organs, sudden changes of temperature, and especially cold, act injuriously. The rule, therefore, established in both civil and military hospitals, that convalescents from fever and acute diseases must be kept warm, should be rigidly enforced."—*Hospital Construction and Management* (John Hopkins).

of the air, the object to be aimed at is, that, in the
wards, it will never fall much below the standard of
the external air. It should not contain more than
0.5 per 1000 volumes of carbonic acid gas, nor such
an amount of organic matter as to be perceptible to
the sense of smell. The breathing of vitiated air will
thus be prevented, and the risk of infection spreading
from the patients to the attendants, or from the
occupants of one ward to those of another, will be
reduced to the minimum, while, at the same time,
the walls, floors, and furniture are less likely to absorb
or retain infectious agents. When a proper volume of
fresh air is continually being mixed with that in a
ward, and a continuous current is ever bearing
away every volatile taint and infectious particle, the
atmosphere is always prevented from becoming
charged with the organisms of disease. A person
entering, or being treated in, a well-ventilated ward,
in contradistinction to a ward in which the air is
charged with the emanations from the sick, may be
compared to a person standing " on a plain across
which a file of men are firing. The chances of escape
are of course better if there are but ten men firing
instead of a hundred."

A healthy adult person requires 3000 cubic feet
of fresh air per hour, in order that the respiratory
impurities may not exceed 0.2 per 1000, or a total
impurity of 0.6 per 1000. If the cubic space per head
be 1000 feet, then the air needs to be changed three
times every hour. If the renewal is affected steadily

and gradually, the cold air is broken up and, mixing with the warm air of the apartment, creates no draught.[1] Dr. de Chaumont found that the smell of organic matter was perceptible in hospitals when the carbonic acid amounted to 0.166 per 1000, so that if 3000 cubic feet be required per hour in health, 4000, or a third more, should be allowed in sickness of an ordinary character, and 6000 in infectious hospitals appears to be necessary.[2] To supply this amount without causing draught, the cubic space must be proportionately increased. Two thousand cubic feet is recommended by almost all authorities for infectious wards. To secure that this amount of air per patient per hour is diffused equally at appropriate temperatures through a hospital ward, various appliances have been used. Mr. Percival Gordon Smith[3] states that he knows instances where it was not possible to maintain a temperature, in enteric or typhoid fever wards, of 42° Fahr., and this with the windows practically closed. He is strongly of opinion, and he is supported by other authorities, that for small hospitals in our climate, powerful stoves or grates are best, and that these if necessary may be supplemented with hot-water pipes. It is not, however, sufficient to trust to windows and fireplaces only for ventilation. Sheringham ventilators, with direct openings through the wall, should be provided between the windows, close to the ceiling, for the admis-

---

[1] Hygiene and Public Health, Louis Parkes, pp. 224 and 226.

[2] Dr. De Chaumont, "Theory of Ventilation," in Proceedings of the Royal Society, and Hygiene and Public Health, p. 146.

[3] Epidemiological Society Transactions, 1882-83, vol. ii.

sion of fresh air. Galton states that the combined area
of these should be at least one square inch for every
100 cubic ft. of space. Outlet shafts should also be
provided, such as Boyle's or Buchan's roof extracting
ventilators. These should be connected with the
wards by enclosed vertical shafts. A shaft should
also be made in the wall alongside the chimney,
dividing at a certain height into two, with an opening
some feet from the ceiling near the corners of the
ward. The heat of the chimney will cause a draught,
which will extract foul air from the angles of a ward,
where it would be likely to remain more or less stagnant. Horizontal trunks are objectionable, as likely
to become receptacles for infected dust. Mr. Percival
Gordon Smith also recommends short flues, passing
directly through the wall, at the level of the floor, one
beneath at the head of each bed, and states that such
openings may have a superficial area of 100 square
inches or more ; and that if provided with sliding
shutters, for occasionally closing them, they are not
in practice found to be inconvenient. Sir Douglas
Galton found that an adequate change of air will not
be satisfactorily obtained in all cases, unless the outlet
or extracting shafts have a sectional area of at least
1 inch to every 50 cubic feet of space in the room,
for the upper floors ; of 1 inch to every 55 cubic feet
in the rooms on the floors below ; and of 1 inch to every
60 cubic feet for rooms on the lower floors, but that this
depends to some extent on the height of the rooms.[1]

---

[1] *Construction of Hospitals*, p. 15.

All openings for the admission of air should be made so as to allow of being easily cleaned, and this should be done every time the ward is disinfected. It would also be desirable that the extracting shafts should be capable of being cleaned and dusted. These also sometimes act as inlets, and they are liable to contain infected dust. In addition, or supplementary to open fireplaces or stoves, it may in some cases be desirable to provide hot-water pipes along the side walls, fixed about 4 inches above the floor. Such an arrangement would permit of cleaning and dusting. At the same time they would be found very desirable on some occasions for warming the air as it filters through the openings already mentioned.

The nurse's room should be placed in the most convenient position for the ward or wards which she is intended to supervise. In small hospitals this room should be regarded as a combination of a ward kitchen or scullery, and nurse's duty-room. It should not be used as a sleeping apartment. It should be provided with a small range or fireplace, with an oven and boiler. The oven would be useful for keeping various articles of food or drink warm. The boiler should be of sufficient capacity to provide sufficient hot water for bathing purposes, as well as for washing and scrubbing. It might in some cases be supplemented by a hot-water geyser, such as that supplied by Doulton of Lambeth, which would provide boiling water in a few minutes at any time of the day or night, should the fire be low or out. The nurse's

room should also be provided with locked cupboards
for the ward crockery, for medicines, and other
articles. A cool larder of small size, with wire gauze
sides, might be conveniently placed outside the
window, opening inside by a hinged door or pane.
In summer weather it should be kept comparatively
cool by being covered with a few folds of clean linen
wrung out of cold water.

Inspection windows should be placed in such positions as will enable the nurse to see into the wards
on both sides. A scullery sink should also be provided.

A cupboard or small closet should, if possible, be
placed in a warm and dry locality, and be of sufficient
size for the storage of all the clean linen and bedclothes required in the ward block.

A separate bath-room, with hot and cold water,
should be provided for wards of six beds and upwards.
For smaller wards, a light bath on noiseless wheels
might be used for two or three rooms. In such cases
a convenient place for the bath is in the verandah, as
may be seen in the designs of the Local Government
Board, see appendix.

In all hospitals, of whatever size, it is very desirable to have an undressing-room, a bath-room, and a
dressing-room so placed as to enable patients to leave
the hospital after taking their final bath, and dressing in non-infected clothing, without returning to the
wards. This might be secured by having a door from
the outside opening into the lobby leading into the

bath-room. If the bath is fixed, it should be placed with one end to the middle of one of the walls, and be free on both sides. The bath, as well as the floor and walls of the bath-room, should be of such material as will not retain infection, and be easily cleansed and disinfected.

A slop sink with hot and cold water, and a water or earth closet, should also be erected for wards of six beds and upwards.

The position of these, as well as of the bath-room, may be seen from the plans. They should always, however, be separated from the wards by a lobby with cross ventilation. The sinks should be fitted with vitrified stoneware sink or receivers. In small hospitals in rural districts, earth closets are preferable. Instead of earth they should be supplied with powdered peat or sawdust, and the contents mixed with petroleum or paraffin, and burnt daily. A small, simple destructor should be erected in connection with the outhouses for the purpose.

Every hospital, of whatever size, should have a plentiful supply of pure water; without this the necessary cleanliness is impossible. One of the first requisites in securing a site for a hospital is the water supply. Dr. Parkes states that from forty to fifty gallons per head per day are often used. He gives the following as being near the quantity required :—

	Gallons daily.
For drinking and cooking, washing kitchen and utensils	2 to 4
For personal washing and general baths	18 ,, 20
For laundry, washing .	5 ,, 6
Washing hospital, utensils, etc. .	3 ,, 6
Water-closets .	10
	38 to 46

According to Mr. Percival Gordon Smith,[1] "the drainage of an infectious hospital differs in no essential point from the drainage of any other building." He states that "it should be clearly understood by the contractor, and by all concerned, that the drains on completion should be tested in lengths by plugging the lower end of each length, and then filling the length of drain with water. If the drain then failed to hold the full quantity of water for a specific time— say four or six hours—it would be evident that means of leakage existed, for which the contractor would be held responsible. The drains should be laid in direct lines, with uniform gradients between the several points, where a change of direction or gradient occurs; and at each of these points means of access to the drains should be provided, either by a manhole or a lamphole, so that the entire system of drains could be inspected with ease at any moment."

The drains should nowhere pass under any part of the building. Every care should be taken by means of ventilation and trapping to prevent the return of

[1] "On the Planning and Construction of Hospitals for Infectious Diseases," *Epidemiological Society Transactions*, vol. ii. 1882-83, p. 157.

sewer gas into any part of the building. If within
easy distance of the sea, the drain should be carried
below low-water mark. In rural districts far from the
sea, the drainage should not be discharged into a river
or a cesspool. If possible, sufficient land should be
acquired to utilise the drainage by irrigation or subsoil filtration, for which also ashes and other garbage
might be utilised.

Every hospital should have a laundry and drying-
room, a mortuary, a disinfecting-room with a tank in
which soiled or infected clothes could be immediately
steeped in a disinfecting fluid, an ashpit, and a destructor for excretions and garbage. In most hospitals an
ambulance shed should be provided. These outhouses
should be erected at least forty feet from the ward
blocks, and as far from the administrative building and
from the wall surrounding the hospital grounds.

Except in very small hospitals the wash-house and
laundry might be in separate apartments. They should
be in size and equipment proportionate to the size of
the hospital and the disinfectant requirements of the
district. The wash-house should have tiled or cemented
floor, non-absorbent walls, and be fitted with wash-
tubs of glazed porcelain or fireclay, as well as with a
boiler and other necessary appliances. The laundry
should have a mangle and ironing table, and shelves
for clothes. The stoves for heating the irons should
be utilised at the same time for heating the drying-
room. In all places, when the weather permits, all
clothes should be dried in the open air.

The mortuary should be of sufficient size, well ventilated, and lighted from the roof. It should have a tiled or cemented floor and walls, so that it can be easily washed with water. It should have a slate bench for coffins, and a table for post-mortem examinations, as well as a sink provided with a water tap. It is desirable to have an inspection window through which friends may be able to have a look at the dead without danger, and the door should be in such a position that the removal of the dead cannot be seen by patients in the wards.

The ashpit should be properly constructed, with cement floor and walls, and roofed, and be of small size. A dry earth closet should be erected in connection with the outhouses for the use of the staff, and in hospitals for one infectious disease it may only be used by the patients, as in Plan A of the Local Government Board.

There is no destructor for burning excretions or garbage suitable for small hospitals, so far as I know, in the market, but I have no doubt the demand would soon create a supply.

It is necessary to devote a separate chapter to the subject of disinfection.

The furniture of isolation wards should be selected with a view to the comfort of the inmates and the cheerfulness of the wards, while at the same time due regard must be paid to economy. The furniture should be of such material as will be least liable to harbour infection, or of such a nature that it can be rendered free of infection with the greatest possible ease. Some articles which entail much trouble and expense in disinfecting may preferably be of cheap material, so that no great loss will be incurred if at any time it be found necessary to destroy them. The wards should have no curtains, carpets, rugs, or mats of any kind.

The bedsteads should be of iron, painted or enamelled, and should be provided with wire, coil, or spring mattresses. The bedsteads should be from 6 feet to 6½ feet long, 3 feet broad, and about 2 feet high. A few cots should be provided for children.

The difficulty of providing a suitable mattress for small Isolation Hospitals is not yet solved.

Miss Florence Nightingale[1] states that "no bedding but the hair mattress has yet been discovered that is fit for hospitals. It does not readily retain miasma, and if it does, heat easily disinfects it. It may be washed. It is not hard to the patient. It saves the objectionable use of a blanket under the patient. There have been repeated objections made to horse-hair on account of the current expense . . . but it is less than is supposed. . . . Straw palliasses are inadmissible—cold, abstract heat, and lessen the chance of recovery." On the other hand, Mr. A. G. Howse[2] states that attempts have been made to disinfect a hair mattress *en masse* by subjecting it to both dry and moist heat at a temperature of 350° Fahr., but that the texture becomes greatly damaged in the process, and

[1] *Notes on Hospitals*, pp. 80-81.
[2] *Treatise on Hygiene and Public Health* (Stevenson and Murphy), vol. i. p. 783.

that if a hair mattress is to be disinfected, it must be pulled entirely to pieces. The expense, according to Miss Florence Nightingale,[1] is not excessive, being about $2\frac{1}{4}$d. to $2\frac{1}{2}$d. per bed, and the loss of hair at $1\frac{1}{2}$d. or 2d.

Dr. Parsons[2] states that horse hair will bear a higher temperature than woollen, cotton, or linen goods —that the process of curling it for stuffing chairs is effected by exposing it to a temperature of 300° Fahr.

Dr. Thorne Thorne[3] states that he found thin horse-hair beds over spring coiled wire mattresses in use in Isolation Hospitals in England, and highly spoken of both by the staff of the hospitals and by the patients, and that, owing to the elasticity of the spring mattress, and consequently the horse-hair beds being less than the usual thickness, they are easily dealt with in an efficient disinfecting stove.

As I have already shown, efficient disinfecting stoves are, owing to their expense, not easily provided for small Isolation Hospitals in this country. It remains, therefore, to be considered what is the best substitute for a hair mattress in small hospitals.

Flock or woollen mattresses must also be cleaned and disinfected by superheated steam. Straw or chaff or husk beds appear to be the only alternatives. A chaff bed is more comfortable than a straw bed, but it is difficult to get clean chaff in sufficient quantity.

[1] *Notes on Hospitals*, p. 80.
[2] Report on Disinfection by Heat, 1884, p. 22.
[3] Supplement to Report of Medical Officer of Local Government Board, 1880, p. 16.

When these are used the tick, when necessary, may be emptied and disinfected and its contents burned. Such beds are in use in many Isolation Hospitals in this country at the present time.

Draw sheets and Mackintoshes should be used to prevent the beds from getting soiled, and render the necessity of destroying them less frequent.

A small square table should be placed beside each bed for the convenience of the patient. These should be painted and varnished or enamelled. Lockers are objectionable. In spite of all care on the part of the nurses they soon become depositories of all kinds of rubbish, and when the ward is to be cleaned out for another disease it is extremely difficult to disinfect lockers properly. There should be one or more similar tables, but of larger size, for the general use of the ward.

A few reclining and other chairs should be provided for convalescents. They should have no stuffing or cushions. In the case of very weak or exhausted patients air cushions may be used.

Other articles of furniture need no description.

Appended is a list of furniture and utensils required in hospital wards.

The administrative building should be comfortably furnished, more or less like a private dwelling intended for persons of the same social grade as those likely to occupy it. The equipment of the kitchen must of course be in excess of the requirements of the number of nurses and officers. It should be sufficiently large and equipped to cope with the work required to provide varied nourishments for the largest number likely at any time to be in the hospital, whether patients or officers.

A medicine chest or press should be provided in the administrative building, in the medical officers' room when such is provided, or in the absence of this room, in the matron's or head nurse's sitting-room. It should contain such medicines as are likely to be required.

LIST of FURNITURE required for ISOLATION HOSPITALS, 2 wards and 8 patients (modified from Dr. Norton Folsom's list).[1]

8 Bedsteads (Iron with spring mattresses)	1 Square table for nurse
10 Hair mattresses or	1 Rocking chair ,,
10 Straw or chaff mattresses	2 Small chairs ,,
10 Pillows	1 Table cover
10 Bolsters	Shirt buttons
16 Quilts	1 Iron bedstead ,,
32 Sheets	1 Washstand ,,
8 Half sheets	Bedclothes ,,
4 Mackintoshes	4 Towels ,,
16 Pillow cases	4 Screens for wards
16 Bolster slips	2 Baskets for soiled linen
4 Air cushions	8 Bedstead card frames
24 Shirts and chemises	2 Clocks
4 Table cloths	2 Thermometers (Ward)
4 Table covers	8 Flower vases
18 Dressing towels	2 Clothes brushes
4 Roller towels	8 Hair brushes
2 Reclining chairs	8 Combs (large)
8 Small wooden chairs	8 Combs (small)
2 Wheeled chairs	2 Nail brushes
8 Small tables	1 Pair scissors
2 Large tables for wards	2 Soap-dishes
	2 Wash-basins

(In Administrative Block: 1 Iron bedstead, 1 Washstand, Bedclothes, 4 Towels)

[1] *Hospital Construction and Management* (John Hopkins), p. 98.

2 Baths
4 Bath sponges
2 Towel stands
2 Looking glasses
4 Hot-water bottles or pans
8 Knives
8 Forks
2 Carving knives and forks
1 Soup ladle
8 Table spoons
8 Tea spoons
1 Corkscrew
8 Plates
8 Small plates
8 Bowls
8 Mugs
2 Medicine cups
2 Measure glasses
8 Cups and saucers
1 Sugar Bowl
1 Cream jug
8 Tumblers
4 Salt cellars
1 Meat dish
1 Soup tureen
1 Coffee pot
1 Tea pot
1 Kettle
1 Sugar box
1 Bread box
4 Trays
2 Water pitchers
4 Bed pans (with covers)

2 Slipper pans
2 Commodes
8 Wash bowls
8 Spit cups
2 Spittoons
1 Chamber pail
1 Dust pan
1 Filler
2 Fenders
2 Pokers
2 Tongs
2 Coal scuttles
2 Shovels
2 Small shovels
4 Candlesticks
4 Lamps
4 Funnels
1 Oil can
1 Lamp filler
4 Shades
2 Match safes
1 Pair lamp scissors
1 Dish Tub
2 Brooms
2 Dust Brushes
6 Dusters
2 Window brushes
2 Mops
2 Dish Mops
2 Scrubbing brushes
2 Glass or earthenware jars for disinfectants

ALLOWANCE of LINEN per BED in chief GENERAL HOSPITALS.   Henry C. Burdett.[1]

MEDICAL WARDS:

4 Sheets
3 Blankets
1 Counterpane
3 Pillow cases

1 Draw sheet
12 Doctor's towels (per ward)
6 Round towels   ,,
4 Table covers   ,,
6 Tea   ,,   ,,

[1] Cottage Hospitals, p. 231.

6 Dusters per ward
6 Shirts   ,,
12 Finger napkins per ward
1 Nightingale cloak
1 Mattress or bed, flock or horse-
    hair
1 Bolster
1 Feather pillow
1 Straw palliasse

NURSES:

1 Quilt each and 2 over
3 Blankets   each
3 Sheets   ,,
2 Pillow cases   ,,
4 Towels   ,,
6 Table cloths   ,,
6 Oil baize toilets   ,,

1 Mattress, horsehair each
1 (Hair) bolster   ,,
1 Feather pillow   ,,

SERVANTS:

1 Quilt each and 2 over
3 Blankets   each
3 Sheets   ,,
2 Pillow cases   ,,
4 Towels   ,,
3 Rollers   ,,
12 Tea cloths   ,,
12 Dusters   ,,
6 Table cloths   ,,
4 Oil baize toilets   ,,
1 Mattress, horsehair ,,
1 Hair bolster   ,,
1 Feather pillow   ,,

CHAPTER VII

DISINFECTION

THE disinfecting process or apparatus provided within the grounds of an Isolation Hospital, as well as the laundry and wash-house, should not only be of sufficient resource to meet the requirements of the hospital, but also of the whole district to which the hospital administers, in so far as the disinfection of infected articles is concerned. The disinfecting-room, as well as the laundry, wash-house, and stores for disinfected washed clothes, should therefore be out of the range of infection from the hospital, and also at a safe distance from the wall surrounding the hospital grounds. Forty feet is believed to be a safe distance from the wards and area wall.

In every Isolation Hospital the disinfection or destruction of excretions, refuse, ward clothing., etc., must be carried out from day to day. The clothes worn by each patient on admission must also be disinfected and stored. Before a patient is discharged he must be thoroughly washed and disinfected, and provided with perfectly clean non-infected clothes.

Every hospital ward must be occasionally emptied of its furniture, beds, and bedding, all of which should be thoroughly disinfected to prevent the accumulation of noxious effluvia by continuous occupancy.[1]

In large hospitals the provision of one or two "fallow" wards has been recommended for this purpose. In such cases one or two wards are always empty or "fallow," and undergoing æration, washing, and disinfection. As soon as this is completed, these wards may be used and other wards emptied for the same purpose. In small hospitals, such as I have under consideration, this plan is not practicable, but probably taking one year with another, the benefit of the "fallow" ward system will be so far secured by the less continuous use of the ward. In small hospitals again, the same wards, the same beds and bedclothes, must be utilised for different infectious diseases at different times, and it will be absolutely necessary, every time a change is made, that the wards and all articles contained in them, as well as their accessories, should be thoroughly disinfected. In addition to all this, the disinfecting and washing resources provided should be able to cope with any extra pressure brought about through infected articles being sent in from any of the dwelling-houses in the surrounding district for disinfection.

[1] Dr. Mahomed noticed that at the fever hospital at Islington, London, patients suffering from scarlet fever recovered better after the wards were washed and purified. There was less sloughing and glandular abscesses. Dr. Howse of Guy's Hospital noticed something similar in his private practice. The first case treated in a room was mild, but the other cases were more severe as the room got saturated with infection.

That, by the adoption of proper means, articles of
clothing, as well as premises and furniture, can be
rendered free from infection and safe for use or
occupation, has been proved beyond a doubt. The
following statement by Dr. Gayton, a gentleman who
perhaps has had more experience in the treatment and
management of infectious diseases than any other now
living, is of importance, as bearing on this question : —
" I have no doubt as to the possibility of rendering a
hospital perfectly free from infection. . . . On 15th
October 1874 1695 cases of smallpox had been received into the Homerton Smallpox Hospital. It was
then decided that scarlet fever patients should be admitted, and accordingly preparations were at once made
for their treatment ; 122 of this class passed through the
wards without any case of smallpox arising. Again,
on the 22nd October 1875, scarlet fever patients were
admitted, 49 in all, without smallpox appearing (no
case of variola had been admitted meanwhile). On
22nd June 1876, the reception of smallpox was recom-
menced—620 being treated during the year without
scarlet fever being diagnosed amongst them. In
1877, 1935 persons suffering from smallpox were admitted, and in 1878, 964. On 15th March enteric
fever cases were received, the total number being
154. Amongst these a case occurred of undoubted
smallpox, but it was admitted from Deptford Hospital,
where smallpox was treated, and it is a matter of

[1] Smallpox and Fever Hospitals Commission, Evidence, Questions 2682,
2690, and 2691.

doubt as to which hospital should be credited with the origin of the attack. At intervals enteric fever, scarlet fever, and smallpox patients occupied the same ward, the same beds, and the same clothing, and without any evil result following. Every single article of clothing was passed through the oven."

In the hospital referred to, however, an expensive disinfecting apparatus is provided. This is not practicable in every hospital in small towns and rural districts. The most reliable apparatus at present provided in this country cannot be purchased for less than £100. The machine requires the most careful attention and skill on the part of the manipulator, and indeed the upkeep does not fall far short of the same amount per annum. Such an expenditure being almost prohibitive in many districts, it remains to be considered what other effective means may be attainable.

The science of disinfection has made considerable progress during the past twenty years. I have already shown how the agents of disease have been demonstrated to be of the nature of micro-organisms, how these have been isolated, made to grow outside the human or animal body on various media, and how it was proved that they retained their power of producing disease under various conditions. The micro-organisms of various diseases have been also subjected to various agents with a view to find out the best practicable and most inexpensive method of destroying their vitality, with a special view of discovering reliable means for the disinfection of infected clothing and other articles.

In 1886, Dr. Parsons of the Local Government Board, in conjunction with Dr. Klein, carried out a series of important experiments, with a view to discover the effect of heat in destroying the micro-organisms of disease. For this purpose they selected of those micro-organisms known to them the hardiest and most resistent to every form of heat, and not only in bacillar or plant form, but also in spore or seed form, in which state the agents of some infections have previously been proved to retain their vitality for a very long time and under very unfavourable conditions. They reasoned that "such arrangements as would be adequate to destroy these 'infections' might be trusted to destroy the potency of infectious matter generally." The following were the infective materials employed: —(1) Blood of a guinea pig, dead of anthrax, containing *bacillus anthracis* without spores—this is the micro-organism which causes woolsorter's disease in human beings; (2) Pure cultivation of the same bacillus in rabbit broth without spores; (3) Cultivation of the same bacillus in gelatine with spores, this being the hardiest form in which this bacillus is known to exist; (4) Cultivation of the bacillus of swine fever in pork broth; (5) Tubercular pus from an abscess in a guinea pig which had been inoculated with tubercle. This contained the bacillus which causes consumption in human beings. Unheated portions of these materials, as well as portions heated to different temperatures, were tested on animals in all cases, in the experiments made.

They first experimented on the effect of dry heat at different temperatures on these materials, and they found that "the spores of the *bacillus anthracis* (the most resistent to every form of heat) lost their vitality, or at anyrate their pathogenic quality (their power of producing disease), after exposure for four hours to a dry temperature a little over the boiling point of water (212-216°), or for an hour to a temperature of 245° Fahr. Non-spore-bearing bacilli of anthrax and of swine fever were rendered inert by exposure for an hour to a temperature of 212-218° Fahr., and even five minutes exposure to this temperature sufficed to destroy the vitality of the former, and impair that of the latter."

They afterwards experimented with boiling water, or water mixed with a very small quantity of common salt ($\frac{1}{2}$ per cent), and they found that a cultivation of *anthrax bacillus* containing spores was rendered inert by five minutes boiling in this very dilute brine. The above results being so satisfactory, it was not considered necessary to repeat the experiments with the less-resisting contagia of swine fever or tubercle.

Experiments were then made with steam at the temperature of boiling water (212° Fahr.). Dr. Parsons states in regard to these: "The results of the above experiments are conclusive as to the destructive power of steam at 212° Fahr. upon all the contagia submitted to its action; in one instance only was there room for suspicion that the disinfection had not been complete—

this was in the case of the highly-resisting spores exposed to steam for five minutes only. On the other hand, the animals inoculated with unheated portions of the same materials all died." . . . "In view of the above satisfactory results, it was not deemed necessary to make any experiments as to the disinfecting powers of steam at higher temperatures or under pressure; its efficiency may be taken for granted."

It will thus be seen that the most persistent or the most tenacious of life of the micro-organisms of disease can be destroyed by a few minutes boiling in water, or by exposure to steam at 212° Fahr. for a few minutes, and that bacilli which do not bear seeds or spores may be rendered inert by exposure for one hour to even a dry heat of 212-218° Fahr. "As none of the infectious diseases for the extirpation of which measures of disinfection are in practice commonly required, are known to depend upon the presence of bacilli in a spore-bearing condition, it may be concluded that, as far as our knowledge goes, their contagia are not likely to retain their activity after being heated for an hour to 220° Fahr."

"Of diseases affecting the human species, anthrax is the only one of which it is established that it is connected with the presence of micro-organisms possessing this persistent form of spores."[1]

Articles infected by any of the infectious diseases prevalent in this country can therefore be disinfected

[1] Extract from Report of Medical Officer of the Local Government Board, 1886: Dr. Parsons, p. 10.

by steam at 212° Fahr., by boiling in water for a few minutes, or by being heated through for an hour in dry heat at 220° Fahr.

The next point to be tested was the rate of penetration of dry heat and steam into bulky articles of non conducting materials, such as pillows, mattresses, etc. Dr. Parsons showed by experiments with various disinfecting apparatus, minutely described in his report, "how difficult it is to secure the penetration of a dry heat sufficient for disinfection into the interior of such articles as a pillow. It was only effected by either employing a very high degree of heat, or by continuing its employment during many hours, the length of exposure compensating for a lower degree of heat." The outside was often scorched before the temperature in the centre of the pillow was sufficient to destroy infection. If heat was applied by means of steam instead of dry heat, he found that the rate of penetration was much in favour of the former. Dr. Parsons came to the conclusion that "it cannot be doubted that to procure the penetration by heat of bulky articles of badly conducting material, high-pressure steam is the agent *par excellence.*"

The following experiment among many others is recorded by Dr. Parsons. "Three yards of flannel previously dried were wrapped round a registering thermometer, making a roll 12 inches long by 4 inches broad. This was placed in the cylinder before mentioned and exposed for ten minutes to steam at 212° [Fahr.] The thermometer registered 212° [Fahr.]

The same roll of flannel, after drying, was placed in a hot-air bath of pots, and exposed for an hour to a temperature of 212° [Fahr.]. The thermometer inside then registered 130° " [Fahr.¹].

Having established the facts that exposure of one hour to dry heat of 220° Fahr., or five minutes exposure to boiling water or steam at 212° Fahr., was sufficient to destroy the infective materials of the infectious diseases treated in Isolation Hospitals, and that steam penetrated with more ease and certainty into the centre of bulky articles, the next point for consideration was the liability to injury of articles disinfected by heat.

Dr. Parsons classified the ways in which injury may occur to various articles, into—(1) scorching, (2) overdrying, (3) fixing of stains, (4) melting of fusible substances, (5) alterations in colour, (6) shrinkage and felting together of woollen materials, (7) wetting. He found that "overdrying can of course only occur when dry heat is employed, wetting only when steam or boiling water is used. Shrinkage takes place more with steam than with dry heat. Scorching is more liable to occur with dry heat than with steam. The other occurrence may happen with heat in either

---

¹ The experiments with steam were made in an improvised apparatus, similar to that employed in Dr. Koch's researches. It consisted of a tin cylinder 14 inches long by 4 inches in diameter, wrapped in felt to diminish loss of heat. In the bottom a hole was made, and by means of corks and a short piece of pipe the cylinder was connected with a tin flask serving as a boiler. A false bottom was fixed in the cylinder just above its base, with a view to the more even distribution of the heat. The upper end of the cylinder was closed with a lid having a hole in the centre, just large enough for the escape of steam, and the introduction of a thermometer. See *Report*, p. 10-11.

moist or dry form." It would appear, however that exposure to temperatures which have been proved to destroy infection does not cause much injury to the majority of articles. Thus white flannel was exposed to steam at 212° for half an hour, and it shrank 5.8 per cent of its length, and got a slight yellowish tinge. On being simply washed it shrank 6.6 per cent, with a very slight change of colour. When boiled in water for half an hour, it shrank 8.4 per cent, and took a dirty yellowish tinge. Gloves, however, were spoiled by exposure for five minutes to steam at the temperature of boiling water, so were shoes and other leather articles. Such articles, however, were not affected by exposure to dry heat of 220°. A felt hat was not altered by exposure for one hour to a dry heat of even 250°. Dr. Parsons states that "the colour and tenacity of white flannel were affected by even a moderate exposure to heat; blankets also were deteriorated in a minor degree. Cotton, black cloth, and silk were little affected by temperatures under 300° . . . Leather will bear a moderate application of dry heat, but is utterly disorganised by a very short exposure to steam." Stains got fixed by a heat of 212° Fahr., whether dry, or steam, or boiling water. According to M. Vallin, however, a washed piece of flannel exposed to a temperature of 230° Fahr. for three hours did not show any difference in colour from another portion not heated, of the same (washed) material. Dr. Russell of Glasgow[1] states in regard to the application of heat

[1] *Glasgow Medical Journal*, Dec. 1884, pp. 404, 405.

for the disinfection of bed and body clothing, and bedding, "Yet it is also true that this disinfectant is the most difficult of all to apply without injury to the textures submitted to it. The material is derived from many sources, animal and vegetable, and varies therefore in the power of resisting heat. The value of the manufactured article depends sometimes on form and elasticity, which may be lost through heat, or on colour, which may be impaired. Every one knows that blankets and other woollen articles cannot be boiled without serious injury. Even the most cautious washing changes gradually the white fleecy new blanket into the yellow, dense, bare, comparatively comfortless old one. With cottons and linens there is no trouble. Wool, hair, and feathers are most troublesome. They all depend for their value and utility upon form and elasticity, which again depend not only upon the hygrometric moisture, but upon the presence of animal fats."

Koch and Wolfhügel [1] were asked to advise the German Government in 1881 as to the efficacy of heat for disinfection. They experimented on the degree of heat necessary for the destruction of various micro-organisms on the same lines as Drs. Parsons and Klein, and they came to the conclusion that bacteria free from spores cannot withstand an exposure of half an hour to a temperature a little over 212° Fahr. in hot air, but that a dry heat of 284° Fahr., continuously applied for three hours, was necessary to

---

[1] *Mittheilungen aus dem Kaiserlichen Gesundheitsamte*, Berlin, 1881.

destroy the spores of bacilli. With steam, however, they found that five minutes exposure at a temperature of 212° Fahr. sufficed to destroy the spores as well as non-spore bearing bacilli. They also found that with steam the heat penetrated into the interior of bulky articles much more quickly than with dry heat. With hot air or dry heat, they found that after an exposure of three or four hours to a temperature of 284° Fahr., small bundles of clothes, pillows, etc., were not disinfected, and that, at that temperature, most materials were more or less injured.

The experiments of Drs. Parsons and Klein therefore confirmed those carried out by Koch and Wolfhügel two years before. Yet in the practical application of heat to disinfection many difficulties are met with. Although articles consisting of hair, cotton, and linen may be subjected to the required degree of any form of heat, the wearing qualities of feather and woollen articles may be impaired in many ways. While steam or boiling destroys leather, dry heat causes more scorching of other articles, and its slow penetration renders its use uncertain and ineffective in the disinfection of bulky bad-conducting material.

Some chemical poisons have also been found suitable for the destruction of the micro-organisms of disease. In the application of these, however, the strength of the solution or vapour used, the duration of its action, the nature of the substance with which the micro-organisms are mixed, as well as the nature of the agents of infection, must be considered. The

M

spores of bacilli are more resistent than the bacilli
themselves to the action of poisons in the same
manner as to the action of heat. The medium in
which micro-organisms may be embedded, or the
substance by which they are surrounded, may exhaust
the destructive power of the poison and prevent it
from reaching them. Corrosive sublimate, which was
found by Koch to be the most active bacterial poison,
is precipitated by albuminous fluids. This renders
the poison insufficient for the disinfection of tubercular
sputum. In non-albuminous fluids, however, Flügge
states that 1 part to 1000 kills even all spores in a
few minutes. Koch also found that a 5 per cent
watery solution of carbolic acid destroyed spores
between the first and second day. He, however,
found "that sulphurous acid gas, even when as concentrated as possible — a degree of concentration
which cannot be attained in practice — only kills
spores imperfectly," and that its action on non-spore
bearing bacteria was uncertain when present in 10 per
cent by volume, owing to its defective penetration, and
to the fact that objects subjected to it must be previously moistened. The same is true, although to a
less extent, of chlorine and bromine gases.

5 per cent of permanganate of potash, and 1
per cent of osmic acid in water destroyed the organisms
on the first day.

The use of dry heat may thus be set aside
as inefficient and unsatisfactory. The remaining agents, such as steam, boiling water, and some

chemical poisons and gases, may be used under varying circumstances. The efficiency of chemical fluids
of known strength is regarded, and must necessarily
be regarded, as uncertain [1] under various conditions.
Their use has also been looked upon more or less as
unnecessary.[2] The proper application of solutions of
carbolic acid or corrosive sublimate of sufficient
strength has not, so far as I am aware, been proved
to be inefficient in destroying the agent of infection in
the case of any of the ordinary infectious diseases prevalent in this country. Common sense is sufficient to
show that masses of solid or semi-solid matter, such
as fæces, or large lumps of dried pus, or smallpox
scales, may be impenetrable to these fluids in the
strengths used, and within the time generally allowed
in ordinary everyday practice. That, however, does
not prove that the small epithelial scales and microscopic dust within which the micro-organisms of infectious diseases must be commonly embedded are not
affected by these agents. These scales and particles
are highly hygroscopic. They easily absorb moisture.
They are softened by water, and being so, the organisms contained within them must, if sufficient time be
allowed, be exposed to the poison dissolved in the
water. Further, these chemical fluids are dissolved in
such strengths as would destroy the most resistent
spores known, although none of the agents of the

---

[1] *Disinfection by Heat*, Dr. Parsons, p. 18.
[2] " On Disinfection," Dr. Russell, *Glasgow Medical Journal*, 1884, pp. 406, 407.

infectious diseases prevalent in this country have yet been proved to propagate by means of spores.

The following extracts from an article by Dr. Russell of Glasgow, and from Reports by Dr. Gayton of the North - Western District Hospital, London, point to the utility of corrosive sublimate solution, at least for some purposes.   Dr. Russell states : " We have during the last ten years washed in the same washing-house over a million of articles of every sort, infected by every variety of contagium known in this country.   Everything has been done exactly as any good housewife would do it, only in a place provided for the purpose, and with ample supply of water and steam, and recently with mechanical aid.   Blankets and woollen articles have not been boiled ; all others have.   The most crucial fact is this, that there has never been a single case, or suspicion of a case, of interchanged disease, *e.g.* of smallpox, appearing in a house from which clothes had been removed on account of scarlet fever or typhus.   In short, I am convinced that in every case the result was obtained for which the operation of washing was undertaken. The only defect is this, *that the washerwomen must handle the articles before disinfection or drowning of the contagia in water, and therefore are occasionally infected."* [1]

Dr. Gayton informs me that before he took charge of the North-Western District Hospital the nurses and assistant nurses were frequently attacked with

[1] *Glasgow Medical Journal*, 1884, vol. ii. p. 409.

typhoid fever. At that time the soiled linen, as soon as it was taken off the beds and patients, was stored in a box in a room off the ward. It was afterwards sorted by the nurse or assistant nurse, and sent to the laundry. The doctor attributed the outbreaks of fever among the staff to the inhalation of infected dust during the process of sorting this soiled linen, and he advised the hospital committee to construct tanks communicating by a shoot with the wards. Into these tanks he put a solution of corrosive sublimate of 1-1000, and gave instructions that all articles of clothing should immediately be steeped in them. Since that time the staff have not suffered from typhoid fever, as the following extracts from his report show : " I have much satisfaction in stating that not a single case of enteric fever has occurred, and the exemption I believe to be mainly owing to your willingness and co-operation in allowing me to have constructed the means whereby all the foul linen is instantly removed from the wards into tanks containing an antiseptic fluid, in which it remains until transferred to the laundry. If the fresh stools of persons suffering from typhoid fever are capable of infecting those in immediate contact, and of which I had conclusive proof some time ago, the placing of linen soaked with discharges in receptacles excluded from air, allowed to remain many hours, and then freely handled for the purpose of sorting, would appear to be a most ready way of communicating the disease." [1] In

[1] Report of Medical Superintendent of North-Western District Hospital, 1884.

1893 Dr. Gayton reports : " One case also was affected by enteric fever undoubtedly contracted in the discharge of her duties, the first since my connection with the hospital."

The steeping of the clothes in pure water might have answered the same purpose in so far as the fixing of dust and the volatile agents of disease is concerned, but in that case the water would be charged with the micro-organisms of disease, and might itself become a medium of infection. If, however, the steeping solution is destructive to bacteria of all kinds, it will answer the double purpose of fixing volatile infection, and destroying at least part of it, while no injury whatsoever results to the articles steeped. At the same time the fixing of stains is prevented, which otherwise might result from boiling.

At the N.W. District Hospital of the Metropolitan Asylums Board all washable articles after being steeped are, according to Dr. Gayton, sent to the laundry and boiled for twenty minutes. All the clothes from the enteric, scarlet fever, and diphtheria wards are then mixed together in the process of washing and drying. The clothes for the different diseases are marked, and when ready are returned to the different departments. There has been no interchange of disease between the wards. In a letter from Dr. Russell, quoted by Dr. Parsons in his report, he states : " For many years I have used no other disinfecting method for washable articles than boiling one quarter to three quarters of an hour by steam with soap and soda. I concluded this was sufficient, because, though we threw smallpox, typhus, enteric, scarlet fever, etc., all into one witches' cauldron, I never heard of any inter-communication or continuity of infection." [1]

From the above it may be seen that reliable means for the disinfection of clothes and other articles may be brought within the reach of even the most isolated rural district. Both in small towns and rural districts, Isolation Hospitals should be erected in more or less isolated localities. A spacious drying or airing green should be provided in proximity to the laundry, within the hospital ground. The laundry should in size and equipment be in excess of the requirements of the hospital. A large boiler should be provided in it, in which all clothes made of cotton or linen should be boiled after they have been steeped in a solution of corrosive sublimate for at least one day. The disinfecting room should be provided with a tank containing a solution of corrosive sublimate 1-1000 for this purpose. This will prevent the fixing of stains, it will lessen the danger to the washerwomen arising from breathing infected dust, and it will at least partially disinfect the clothes. The disinfecting chamber should also be provided with a cheap but reliable apparatus for the submission of such articles as cannot be boiled to the action of steam. An apparatus on the model of the cylinder used by Koch, but correspondingly larger, might be constructed for a

---

[1] *Disinfection by Heat*, Dr. Parsons, p. 10.

small amount. Flügge [1] describes a very simple and
cheap apparatus used in Gottingen, and which acted
extremely well. It is made in two sizes : one for the
disinfection of portions of clothing or linen, and costs
£7 : 10s. ; the other is of sufficient size for the disinfection of mattresses, and costs £13. Dr. Flügge
considers that such machines should be round, or
nearly round, to provide against imperfect disinfection
in corners. He also states that it was definitely
proved by numerous experiments in Koch's laboratory,
and confirmed by Wolff, that very large objects, such
as balls of twenty-two blankets, are completely disinfected by exposure for one to two hours to a current
of steam at 212° Fahr. The time of exposure to this
temperature must be reckoned from the time that the
steam issues from the aperture provided for its exit
at a temperature of 212° Fahr. It is necessary to
have a current of steam passing through the clothes.
If this is attained it is not necessary to have steam
under pressure to act as an efficient germicide or for
the penetration of bulky articles. Dr. Parsons has
pointed out that when steam enters into the interstices
of a cold body it undergoes condensation in imparting
its latent heat to the body,[2] that when condensed it
occupies but a small portion of its former space, and
that in this way a series of successive vacua are formed
into which a fresh supply of steam enters until the
whole mass is penetrated. If the materials are only

[1] *Micro-organisms,* by Dr. C. Flügge, pp. 665, 773.
[2] *Report on Disinfection by Heat,* p. 18.

moistened by steam they can be dried in a short time. It is only where they are soaked with water from the condensation of the steam on the sides or top of an apparatus that soakage is liable to occur. This can be prevented by simple contrivances described by Flügge[1] in his description of the apparatus used in Gottingen. The subsequent drying of articles exposed to steam is not such a pressing question in small communities as in large cities, where spacious drying greens are not so easily obtained, and loss of time more considered. For such communities as can afford to provide an apparatus where superheated steam under pressure is used, such as the Washington - Lyons patent apparatus, there appears to be no doubt whatever of its superiority. The most delicate fabrics, as well as the most bulky, can be disinfected with the greatest certainty, and with the least injury, trouble, or loss of time. This apparatus, however, is at present so expensive that it cannot be purchased by small communities. Where no steam apparatus of an efficient nature is provided, disinfection of such articles as mattresses and blankets, which cannot be boiled, can be carried out, although with greater trouble and loss of time. Hair mattresses should be opened and the ticks and hair soaked in corrosive sublimate solution, and subsequently boiled in water. The blankets should also be soaked in the same solution and thoroughly washed. After washing they should be well dried and exposed to air and sunlight for some

---

[1] Flügge on *Micro-organisms*, pp. 774, 775.

days. To enable this to be carried out without interfering with the use of the hospital, an extra supply of blankets and mattresses may be required. Thorough washing without boiling has been found sufficient for disinfection, by Dr. Russell of Glasgow, for blankets and woollen articles.

Boots and other leather articles, as well as hats, should be exposed to dry heat. Ordinary boots, as made in this country, will not stand long soaking in corrosive sublimate solution. In the absence of an efficient disinfecting apparatus, carpets and stuffed furniture should be thoroughly beaten and exposed for some days to the air and sunlight. All wooden and other articles of furniture and household use should be washed with corrosive sublimate solution and afterwards scrubbed with soap and water. The same process applies to walls, windows, floor, and doors.

Appended is a copy of the instructions for disinfection issued by Dr. Collie of the Homerton Fever Hospital.

As soon as a case of infectious disease is removed to a hospital it is the duty of the local authority to see that the house furniture and clothing are thoroughly disinfected. This requires much care and close supervision on the part of the officers appointed to see it carried out. Many difficulties are met with, more particularly in the case of the poor. A large family, inhabiting perhaps one or two apartments, without any other place of abode, cannot well be turned out on the roadside until the house is thoroughly gutted of its

contents and made perfectly safe for use. The clothes
they wear should be disinfected, as well as the bedclothes in which they sleep. From what I have
already stated it will be seen that this is a process
which will take many hours, perhaps a day or two, to
complete. If, however, as much of the infected
clothing as could be spared be immediately conveyed
to the hospital disinfecting chamber and laundry, and
as much of the furniture, etc., as could be spared be
washed and turned out of doors for exposure to the
air, the infected apartment or house could be gradually disinfected without excessive hardship or inconvenience. With the assistance of one or two women
the Sanitary Inspector could see the work completed
within a reasonable time, and without undue friction.

INSTRUCTIONS for DISINFECTION of HOMERTON FEVER
HOSPITAL after being used as a Smallpox Hospital.

Wards, etc.—1. Wards, including nurses' sitting-rooms, sculleries,
bath-rooms, closets, lavatories, linen cupboards, and linen shoots, to
have their windows, doors, chimneys, and ventilators completely
closed, and to be emptied of all movables, excepting the blinds
and the windows.

2. Then burn sulphur in an atmosphere of steam for forty-eight
hours, so as to permeate all the above mentioned places.

3. The windows having been entirely taken out, the wards and
adjoining rooms are to be exposed for fourteen days to wind and
weather.

4. Remove wash from ceiling, and then whitewash it. Wash
well down the walls. All woodwork to be thoroughly washed with
soap and water, and the floors to be well scrubbed. The walls and
woodwork should be washed well, at least twice. After the paint-

ing, the floors should be well scrubbed three times, at intervals of three days. The hoppers and ventilators should be specially looked to.

5. Now fix the windows, new sash-lines having been supplied, and commence painting, which should consist of at least three coats.

Furniture.—1. The venetian blinds to be taken to pieces, and well washed with soap and water, dried in the open air, and re-painted, new tapes and cords being supplied.

2. Bedsteads to be taken to pieces, and well washed and re-painted. Sacking to be washed and dried in the open air.

3. Beds, pillows, and bolsters to be emptied, and the feathers to be washed with super-heated steam at a temperature of $300°$ for twenty minutes, and dried for three hours at a temperature of $250°$. All the linen, ticks, blankets, counterpanes, draw sheets, shirts, chemises, night-dresses, towels, handkerchiefs, nightingales, and squares, to be steeped in fresh water one week, the water to be three times changed, then to be boiled in different waters twice, and afterwards washed and dried in the open air, but the blankets and nightingales are to be subjected to a dry heat of $230°$, and not boiled.

4. Night stools, bed-tables, chairs, bed-boards, tables, cupboards, presses, and racks to be exposed in the open air for fourteen days, then well scrubbed with soap and warm water, roughly wiped, and left to dry in the open air.

5. The seats of the night stools and the seats of all the water-closets to be planed.

6. All sinks, sluice-pans, and closets to be flushed.

7. Crockery and medicine bottles to be thoroughly washed in boiling water. The feeders in particular to be boiled, and the spouts cleaned out with a brush.

8. All brooms, brushes, scrubbing-brushes, knee-rests, flannels, dusters, and sponges used in the smallpox period, and any tow, lint, or cotton-wool which may have been in the wards, to be destroyed; also all mats, and oilcloths, and carpets in the wards, or the sitting-rooms of the nurses.

9. Slop-pails and tubs to be steeped in fresh water for twenty-four hours, exposed to wind and rain, washed in warm water and soap, and dried in the open air and re-painted; all rags and cloths used in connection with these to be destroyed.

Books, papers, toys, etc., to be destroyed or sent to a smallpox hospital.

Clothing.—1. Men's and women's underclothing to be steeped in fresh water for a week, twice changed, then washed and dried in the open air. It would be desirable, however, to send these things to a smallpox hospital for use, as they do not stand boiling well.

2. Boots and shoes used by smallpox patients to be sent to a smallpox hospital or destroyed.

3. So far as possible the clothing of officers, nurses, and servants, worn in a smallpox ward, to be steeped in fresh water for a week, then washed, boiled, and dried in the open air. Clothing so worn which cannot be so treated to be destroyed.

Receiving Rooms.—1. To be treated in the same way as the wards, but new baths to be supplied.

2. New carrying chairs to be supplied.

General.—1. All the furniture in the Asylum, other than that in the wards, to be cleaned in the ordinary way.

2. The rooms and offices of officers, nurses, and servants, where paper exists, to be re-papered after the removal of old paper ; wood-work to be washed and re-painted, and in other respects cleaned in the ordinary way.

3. Carpets and curtains, beds, bedding, sheets, linen generally, and blankets, and all furniture in officers' and servants' rooms, to be thoroughly cleaned.

4. All rooms, passages, corridors, pantries, store-rooms, linen-rooms, stairs and staircases, water-closets, lavatories, and bath-rooms, in the administrative departments, to be well washed as to the wood and stone-work, and the wood-work to be re-painted, and white or lime-washed, where such was the case before.

5. Mortuary and post-mortem room to be fumigated with sulphur and steam, well washed down, exposed to the atmosphere, and re-painted.

Kitchen.—Well washed, exposed to the atmosphere, lime-whited and re-painted. All utensils employed there to be thoroughly cleaned.

Steward's Department.—Store-rooms and offices to be thoroughly cleaned in the ordinary way.

Matron's Department, Linen and Store-rooms.—Simple cleaning.

Laundry.—Simple cleaning.

Disinfecting Chambers.—Simple cleaning.

Patients' Clothes Rooms.—Wooden framework to be taken out

and burnt, to be thoroughly fumigated with sulphur and steam, exposed to the air for fourteen days, well washed, and wood-work re-painted.

Dust holes and Dust shoots.—To be cleaned.

Roads.—Fresh gravel.

In a word, whatever has been in contact with smallpox must, if possible, be cleaned; and what cannot be cleaned, must be got rid of.

CHAPTER VIII

REMARKS ON HOSPITAL MANAGEMENT—AMBULANCES

THE efficient management of Isolation Hospitals, with due regard to economy, is a matter of considerable difficulty. Every hospital, however small, should be under the management of a committee who should be responsible to the Local Authority for the expenditure incurred in its upkeep and maintenance.

Scattered communities isolated from populous centres suffer less from outbreaks of infectious disease. In such places a hospital might be empty for a considerable time. In small towns, and even rural districts in frequent communication with towns, cases of infectious diseases of one kind or another keep breaking out continually. The question of providing a temporary or permanent staff for Isolation Hospitals is therefore worthy of consideration.

In England, patients, if able to pay, may be charged by the Local Authority for medical attendance and maintenance during the time they are isolated in their hospitals. In Scotland, however, the Local Authority has no such power. The expense must be levied on the ratepayers as a whole. In many places in Eng-

land a medical practitioner continues his attendance
on his patient after his removal to an Isolation Hospital, and receives his usual fees.   In small towns or
rural districts where an Isolation Hospital is not sufficiently large to require the services of a resident
medical officer, there is much to be said in favour of
this system.   If the medical officer of health is also
the sole hospital doctor, and is also allowed to practise
in the district, other medical men may hesitate to
advise their patients to go into the hospital.   By doing
so they incur some pecuniary loss, and they increase
the influence of a rival practitioner.   Patients also
naturally follow the advice of their own family
physician, and would prefer to continue under his care
in the hospital as well as at home.   If medical
practitioners resident in a district are allowed to
attend on their own patients in a hospital, there is not
the same necessity of appointing a medical officer at a
fixed salary.   Under such a system practitioners are
paid only when their services are required, whereas
by appointing one man with a fixed salary, he
receives payment whether the hospital is used or not.
It would be very undesirable to have a large number
of medical men visiting a hospital daily.   That, however, would not occur.   Where a large number of
medical men are to be found, there must also be a
large population.   A large population would require a
hospital of such a size as would take up the whole
time of a resident medical officer.

The medical officer of health should advise the

Local Authority in regard to the general management of Isolation Hospitals. He should see that all necessary precautions are carried out with a view to prevent the spread of infection either in or from the hospital. No patient should be discharged except by his orders. He should frame regulations for the admission and discharge of patients, the guidance of nurses and other officers, the admission of visitors, and for the general working of the hospital. He should have entire control under the Local Authority over all measures that may be taken for the protection of the patients and the public.

The visiting physicians, on the other hand, should have entire control over the cases under their charge in so far as the ordering of suitable diet and medicines or other means of treatment are concerned.

" In every hospital a book should be kept containing information as to the circumstances of patients admitted,—the facts to be recorded in it being quite distinct from those relating to the history, symptoms, and course of disease,—and to the treatment adopted. The absence of such an 'admission book' has been found, during the course of this inquiry, to have caused considerable inconvenience to sanitary authorities and their officers when desiring to collect information concerning the usefulness or otherwise of their hospital." Dr. Thorne Thorne suggests the following headings for such a book, and states that for some hospitals part of the information would be unnecessary, while for others additional headings might be required.

N

Number.	Date of Admission.	Names.	Age.	Sex.	Address.	Occupation.	Disease.	Medical Attendant.	Date of Discharge.	Death.	Repaid on behalf of Patient. Amount.	Source of Payment.	Remarks.
1.													
2.													
3.													
4.													
5.													
6.													

In some hospitals, according to their size and requirements, one, two, or more nurses should be permanently employed, while in others this would not be found necessary. When a nurse is not permanently employed, an intelligent woman might be allowed a small remuneration to keep the hospital fired and aired, and always in readiness for the reception of patients. Arrangements might be made for her residence in the administrative building. She might, if necessary, attend to a patient until the arrival of a nurse, if such an emergency occurred. During the time patients are under treatment she might do the cooking, and otherwise render assistance.

In some cases a man and his wife might live rent free in the administrative building, the husband being allowed to earn his living otherwise, when the hospital is not in use, but to act as ambulance driver, or to render such assistance as might be found necessary at reasonable wages, when the hospital was occupied.

It is of great importance that provision should be made whereby the services of one or two nurses might be obtained at any time with the least possible delay. In large counties a permanent staff of a few nurses might be always engaged in one hospital or another. It might suit some districts or counties to pay a retaining fee to a nurse's institute in one of the large towns to secure the services of one or more nurses when required. Ladies in several parts of Scotland have started rural nursing associations for the supply of district nurses. One of these, at least, has in contemplation the supply of nurses to Isolation Hospitals in the district when necessary. In engaging nurses for infectious disease, it is generally desirable to have them protected by a previous attack. In the case of smallpox, revaccination, properly done, is sufficient protection. In diphtheria one attack does not protect a person from the disease. Adults, however, are not so liable to suffer as young persons, and, with proper precaution, the danger of infection is not so great as with some other diseases.

That there is considerable danger of infection being conveyed from one person to another by a third party may be seen from the following facts :—A friend of mine, a general practitioner in a rural district, informed me that one day he had in the morning to visit a person suffering from measles. There were symptoms of lung complication, and he had to make a minute examination of his patient, and come into close personal contact with him. He immediately afterwards drove a distance of seven miles, and when passing a lonely cottage by the roadside was called in to see another case, a child, and here also had to make a minute examination of his patient.   Between ten and eleven days afterwards, when measles had time to incubate, this last case was taken ill with that disease.   The case was isolated in a lonely house in an out-of-the-way part of the district, and had no communication with any infected family. The doctor made a thorough inquiry and had to come to the conclusion that he was the carrier of the infection himself.

It is therefore essential that every care should be taken to prevent an Isolation Hospital from becoming a centre from which diseases may spread.   For this purpose it is necessary to have rules for the guidance of the nurses and others.   In towns where infectious hospitals are often built in thickly populated localities, extreme care is required.   In the North-Western District Hospital of the Metropolitan Asylums Board, Dr. Gayton informs me that no one is permitted to pass the gates (nurses or others) unless furnished with a certificate signed by the bath attendant setting forth that she "has had a bath and changed her hospital uniform for her private clothing."   To further ensure the rule being carried out, a room has been constructed furnished with lock-up cupboards—one appropriated to each member of the staff—where all private articles of wearing apparel must be kept.   This apartment being close to the bath-rooms, the plan of procedure is rendered very simple. though doubtless somewhat irk-

some, and is as follows:—The bath being taken, and
the official uniform left in charge of the attendant, the
nurse passes from this room to the adjoining one containing her out-door clothing, and, being dressed, emerges
from a door at the other end into the grounds of the
hospital. The order of things being, of course, re-
versed upon her return to duty. (Report, 1884, p. 7.)

At the Belvidere Hospital, Glasgow, this is not
done, and Dr. Russell informed me that he was not able
to trace infection at any time to the nurses. The Belvidere Hospital is, however, fairly isolated. Each
ward block is entirely disconnected from the rest of
the building. There are not even covered corridors.
The nurses are often in the open air. They are provided with overcoats, and when going off duty they
simply take off their hospital uniform and wash themselves. In small towns and rural districts this would
probably be found quite sufficient.

"In the Blegdam Hospital, Copenhagen, the
central portion of the front administration block contains two bath-rooms for the visiting medical staff.
Each bath-room has on each side a dressing-room, one
communicating with the entrance hall, the other with
a lobby leading out to the wards. The medical officer
on arriving leaves his ordinary clothes in the front
dressing-room, and passing through the bath-room
assumes his hospital garments. On returning he re-
verses the process, and in addition takes a bath before
putting on his ordinary clothes." [1]

[1] Burdett, *Hospitals and Asylums of the World*, vol. iv. pp. 283, 284.

Visitors to Isolation Hospitals may also convey infection. The following are the rules regulating the visiting of patients in the Hospitals of the Metropolitan Asylums District Board : [1]—

1. The visiting of patients in these hospitals is limited to the nearest relatives and intimate friends of patients dangerously ill. One visitor will be allowed to each of such patients. Such visits can only be made with the permission of the medical superintendent, and will be limited in duration to a quarter of an hour, except in very urgent cases, when two visitors will be allowed, and the duration of the visits may be extended.

2. Notice will be sent to the nearest known relatives or intimate friends of patients dangerously ill, with an intimation that they may be visited. Such notice will be accompanied by a copy of the regulations under which visits can be made.

3. A list of patients dangerously ill will be sent daily at one o'clock by the medical superintendent to the gate porter, to enable him to answer inquiries.

4. Visitors are warned that they run great risk in entering the hospitals. No one should attempt to enter the wards of a smallpox hospital without having been previously properly revaccinated, and if he lives in a house where smallpox has occurred, he is urged at once to apply to the public vaccinator (whose address can be obtained from any of the parish officers) in order that the remainder of the occupiers of such house may be vaccinated.

[1] Smallpox and Fever Hospitals Commission Report, p. 380.

5. Visitors are advised (*a*) not to enter any of the wards when in a weak state of health or in an exhausted condition. (*b*) To partake of food before entering the hospital. (*c*) To avoid touching the patient or exposing themselves to his breath or to the emanations from his skin. (*d*) To sit on a chair at the bedside at some little distance from the patient, and not to handle the bedclothes.

6. Visitors will be required to wear a wrapper (which will be provided at the hospital) to cover their dress while in the wards, and to wash their hands and face with carbolic soap and water before leaving the hospital, or to use some other mode of disinfection at the discretion of the medical superintendent.

7. Visitors are strongly urged not to enter any omnibus, tramcar, or any other public conveyance immediately after leaving the hospital.

On the admission of a patient to any of the hospitals of the Metropolitan Asylums District Board,[1] a letter will be sent to the nearest known relative or friend, setting forth the state of the patient. Should any serious change for the worse take place, a letter will be sent daily to the relative or friend, stating how the patient is progressing, which letter will be continued until the patient is in such a condition as to render further communication unnecessary. But should the patient become dangerously ill, notice will be sent to the nearest known relative or intimate friend that the patient may be visited, and, at the discretion of the

[1] Metropolitan Asylums District Board Regulations, 18th June 1887.

medical superintendent, arrangements may be made
for the conveyance of the visitor to and from the
hospital.   Inquiries as to the condition of the patients
must be made in writing to the medical superintendent, who will reply by return of post.   It is very
undesirable that friends of patients should personally
make inquiries at the hospital.

In the Belvidere Hospital, Glasgow, a large room
is provided with seats near the entrance gate. Visitors
enter this room from the outside at a stated hour every
day.   At a considerable height above the floor there
are openings numbered to correspond with each ward.
The nurse of each ward comes to this opening from
the outside, and gives information regarding any
patients in her ward.

The appearance of the building and the manner in
which the hospital and grounds are kept are of great
importance in creating a kindly feeling towards it in
the neighbourhood.   In many places very little attention is paid to this.   In rural districts too much is left
to nature.   The purity of the outside air will not,
however, make up for unhealthiness caused by overcrowding or damp walls, or insufficient means for
ventilation or want of cleanliness inside an hospital,
or dirt and slovenliness without.

It is quite as necessary to provide healthy, attractive
buildings in the country as it is in towns.

Overcrowding, poverty, and filth will act as a *nidus*
for typhus as certain in a lonely cottage exposed to the
purifying Atlantic breeze as in the dirtiest slums of a

town.[1] Typhoid fever[2] will lurk about a country house of defective construction with the same certainty as in a town dwelling. A crust[3] of smallpox will retain its infecting qualities for years while lodging in the walls of a country house. Defective ventilation, damp walls, soil saturated with organic matter, have the same evil effect in town and country. As Sir John Simon stated, "that which makes the healthiest house makes the healthiest hospital ; the same fastidious and universal cleanliness, the same never-ceasing vigilance against the thousand forms in which dirt may disguise itself in air and soil and water, in walls and floors and ceiling, in dress and bedding and furniture, in pots and pans and pails, in sinks and drains and dust bins. It is the same principle of management, but with immeasurably greater vigilance and skill, for the establishment which has to be kept in such exquisite perfection of cleanliness is an establishment which never rests from fouling itself, nor are there any products of its foulness, not even the least odourless of such products, which ought not to be regarded as a poison."[4]

In many parts of England much attention is paid to the necessity of making Isolation Hospitals attractive in appearance as well as healthy in design, construction, and management.

The little hospital at Ealing is one example of this.

[1] Typhus fever is frequently met with in lonely cottages in Skye and other places in the Highlands of Scotland.

[2] Cases of typhoid in country houses already given.

[3] The case of smallpox recorded by Dr. Carpenter occurred in Mull.

[4] Report of Medical Officer to the Privy Council, 1863.

Not only are the buildings attractive in appearance and well kept, but the grounds around them are beautifully laid out, planted with a profusion of trees, shrubs, and flowers, and kept in perfect order. Even the name by which the building is known, as I have endeavoured to explain in the preface, is of importance in contributing to the success of the institution.

An ambulance carriage or carriages should be provided, according to the requirements of the district, for the conveyance of patients to Isolation Hospitals. Such an ambulance is a necessary appendage to every Isolation Hospital. Persons suffering from infectious diseases are seldom able to walk to a hospital, and even if able, it is very undesirable that they should do so, as they might communicate the disease to such as might meet them on the way. It is also very undesirable that persons suffering from infectious diseases should be carried to a hospital in any private carriage, as such carriages are not easily disinfected.

According to the Public Health Acts, persons suffering from infectious diseases are prohibited from exposing themselves either on streets, roads, or public places, or to travel in any public conveyance. It is, therefore, absolutely necessary for Local Authorities to provide properly constructed ambulances for the conveyance of patients to their hospitals. An ambulance carriage should be so constructed as to enable the patient, if necessary, to lie down. It should be of sufficient size to enable a nurse to accompany the patient. It should be constructed of light but durable material, and of such a nature as to prevent the absorption of infectious organisms. The inside should be plainly furnished, so as to facilitate thorough disinfection each time the carriage is used. The berth occupied by the patient should be a combination of a bed and stretcher, so that it may be carried if necessary into the house before the patient is disturbed, and into the ward where he is to be isolated and put to bed.

There should be a seat for the nurse inside the carriage, a speaking tube to communicate with the the driver, and a small locker in which stimulants and restoratives could be kept. The ambulance should also be properly ventilated, and provided with subdued light. It should have smooth elastic springs, and the wheels should be provided with rubber tyres. It is also very desirable that the ambulance in country districts, where the roads are narrow, with sharp turns, should be so constructed as to suit such contingencies.

Ambulance carriages of the above description are made and sold at reasonable prices by different firms in this country.

The success of Isolation Hospitals in preventing the spread of infectious disease depends to a great extent on the rapidity with which patients are removed from their houses after the disease is declared to be infectious. For this purpose an ambulance should be always in readiness to start within the shortest possible time. In London the organisation for the removal of patients is carried to great perfection. Three ambulance stations, fully equipped with horses, drivers, nurses, helpers, telephone clerk, and superintendent, are provided by the Metropolitan Asylums Board. Applications for admission are generally made by patients or Local Authorities to the central office of the Board. Within three minutes of a summons being received at the ambulance station from the central office for the removal of a patient the ambulance is started on the journey.

The following are some of the regulations of the Metropolitan Asylums Board for the removal of persons suffering from infectious diseases :—

(1) APPLY—On Week Days, between 9 A.M. and 8 P.M., to the Chief Offices :—

Postal Address : Norfolk House, Norfolk Street, Strand.
Telegraphic Address : Asylums Board, London.
Telephone Number : 2587.

N.B.—Applications in the latter part of the day must be dispatched in time to reach the Offices before 8 P.M.

At Night between 8 P.M. and 9 A.M., and on Sundays, Christmas Day, and Good Friday, to the Ambulance Stations :—

Eastern Ambulance Station, Brooksby's Walk, Homerton, N.E.
South-Eastern Ambulance Station, New Cross Road (near Old Kent Road Railway Station), S.E.
Western Ambulance Station, Seagrave Road, Fulham, S.W.

Removal to the Board's Hospitals

(a) ONLY persons suffering from SMALLPOX, FEVERS, or DIPHTHERIA are ADMITTED into the BOARD'S HOSPITALS.

(b) Every application must state the name, age, and full address of the patient, from what disease suffering, and in cases of fever, the particular kind of fever ; and also the name of the person making the application.

(c) Unless a Medical Certificate be handed to the Ambulance Nurse the patient will not be removed.

(d) Patients should leave all valuables, money, etc., and all outside clothing at home, should wear body linen only, and be wrapped in the blankets provided for the purpose.

(c) The Ambulance Nurse will leave, at the house from which the patient is removed, a notice stating the hospital to which the patient is to be taken, and a copy of the regulations as to visiting, etc.

Conveyance to other Places

(a) PERSONS suffering from ANY DANGEROUS INFECTIOUS DISEASE may be CONVEYED by Ambulance to PLACES OTHER THAN THE BOARD'S HOSPITALS.

N.B.—Dangerous Infectious Diseases include the following :—Smallpox, Cholera, Diphtheria, Membranous Croup, Erysipelas, Scarlatina or Scarlet Fever, Typhus, Typhoid, Enteric, Relapsing, Continued, and Puerperal Fevers, and Measles.

(b) Every application for an Ambulance must state :—
    (i.) Name, sex, and age of patient.
    (ii.) Description of disease, and, in the case of fever, the particular kind of fever.
    (iii.) Full address *from* which the patient is to be conveyed.
    (iv.) Full address *to* which the patient is to be conveyed.

(c) The patient must be provided with a Medical Certificate of the nature of the disease, to be handed to the driver of the Ambulance.

(d) The charge for the hire of the Ambulance, including (when the patient is over ten years of age) the services of a male attendant, is 5s. This amount must be paid to the driver, who will give an official receipt for the same.

(e) One person only will be allowed to accompany the patient, and such person may be conveyed back to the place from which the patient was conveyed. If desired, a nurse will be supplied at an additional charge of 2s. 6d. for her services.

(f) The Ambulances may be sent outside the Metropolitan district only by special sanction of the Ambulance Committee or of the Clerk to the Board, and in such cases an extra charge will be made of 1s. for every mile outside the Metropolitan area.

The drivers of the Board's Ambulance are not allowed to loiter on their journeys or to stop for refreshments, on pain of instant dismissal. It is particularly requested that any breach of this regulation, or any neglect or incivility on the part of the drivers, nurses, or attendants may be immediately reported to the undersigned.

The servants of the Board are forbidden to accept any gratuities or refreshments.

BY ORDER.

Dated 25th January 1892.

> N.B.—*By Section 70 of the "Public Health (London) Act, 1891," it is enacted that*—"*It shall not be lawful for any owner or*
> "*driver of a public conveyance knowingly to convey, or for any*
> "*other person knowingly to place in any public conveyance, a*
> "*person suffering from any dangerous infectious disease, or for*
> "*a person suffering from any such disease to enter any public*
> "*conveyance, and if he does so he shall be liable to a fine not*
> "*exceeding TEN POUNDS. . . .*"

(2) NOTICE OF REMOVAL OF PATIENT TO HOSPITAL

Date   189

To the Nearest Relative or Friend of the undermentioned Sick Person.

(Name of Patient)

(Address)

will be removed to the NORTH WESTERN HOSPITAL, situate at HAVERSTOCK HILL, HAMPSTEAD, N.W. (or to Hospital Ships, etc., according to circumstances).

On the other side is a copy of the Regulations as to the furnishing of information relative to the condition of patients, and as to the visiting of patients at such Hospital, to which your serious attention is requested.

> IMPORTANT.—*All clothes used by the above named patient, before his (or her) removal to Hospital, must be carefully disinfected. To ensure this being done apply to the Local Sanitary Authority.*

COUNTERFOIL. (225)

METROPOLITAN ASYLUMS BOARD,

EASTERN AMBULANCE STATION,
HOMERTON,
HACKNEY,

189

Name

Residing at

Chargeable to

This Note to be left at the Hospital with the Patient.

(225)

METROPOLITAN ASYLUMS BOARD,

EASTERN AMBULANCE STATION,
HOMERTON,
HACKNEY,

........Fever 189

Name

Removed from

Name and Address of nearest relative or friend, or of a resident in absence house.

Chargeable to

Superintendent.

DRIVER'S AND NURSE'S NOTE.

(225)

METROPOLITAN ASYLUMS BOARD,

EASTERN AMBULANCE STATION,
HOMERTON,
HACKNEY,

189

Name

Residing at

Chargeable to

Nurse

Driver

Patient recd. by   at

Ambulance arrived...o'clock.

Ambulance left......o'clock.

Gate Porter.

In for Nurse...o'clock.
Out with do...o'clock.

(4)

COUNTERFOIL. (225)

METROPOLITAN ASYLUMS BOARD,

EASTERN AMBULANCE STATION,
HOMERTON,
HACKNEY,

189

Name

Residing at

Chargeable to

This Note to be left at the Hospital with the Patient.

(225)

METROPOLITAN ASYLUMS BOARD,

EASTERN AMBULANCE STATION,
HOMERTON,
HACKNEY,

189

Name

Removed from

Name and Address of nearest relative or friend, or of a resident in above house.

Chargeable to

Superintendent.

DRIVER'S AND NURSE'S NOTE.

(225)

METROPOLITAN ASYLUMS BOARD,

EASTERN AMBULANCE STATION,
HOMERTON,
HACKNEY,

189

Smallpox

Name

Residing at

Chargeable to

Nurse

Driver

Patient recd. by at

Ambulance arrived...o'clock.

Ambulance left ......o'clock. } In for Nurse...o'clock.

Gate Porter. } Out with do....o'clock.

(5) Directions to Nurses

To take charge of the Ambulance, giving any necessary directions to the driver.

To ride inside the Ambulance with the patient, administering stimulant if necessary. If the case be severe, to order the driver to proceed at a walking pace.

To wear the uniform provided by the Managers. To see that the Ambulance is disinfected at Potter's Ferry after delivering up the patient.

To see that no stoppage of the Ambulance takes place on the journey. Any breach of this direction will lead to instant dismissal.

(6) Directions to Drivers

The driver of the Ambulance, on receiving his instructions, will go to the address given him, take up the patient, and proceed with due care and diligence to   , and there deliver up the patient to the satisfaction of the nurse ; after the Ambulance has been disinfected he will forthwith return to the Depot, without stopping or calling at any place whatever, on pain of instant dismissal.

The police have been instructed, and will in all cases report to this office any one loitering or stopping on the road, except for the purpose of taking up or discharging the patient.

(7) Instructions to Drivers of Ambulances

Removal of Patients to other places than the Managers' Hospitals

The patient must be provided with a medical certificate of the nature of his or her disease, and such certificate must be handed to the Ambulance Driver, and be by him delivered, on his return to the Ambulance Station, to the Station Superintendent.

The Driver is to obtain the amount charged for the hire of the Ambulance and for the services of the nurse (when provided), to give an official receipt for the same, and to hand the amount over to the Station Superintendent.

O

Drivers are not allowed to loiter on their journeys or to stop for refreshments, on pain of instant dismissal.

The servants of the Board are forbidden to accept any gratuities or refreshments.

(8) When a driver of an Ambulance is sent to remove a patient to any other hospital than one of the Managers' hospitals, he is to inquire, before the patient enters the Ambulance, whether arrangements have been made for the reception of the patient at the hospital to which he or she is to be removed, and if he finds that no such arrangements have been made, he is to suggest to the patient's friends that, as there may not be room at the hospital, the Ambulance should be again applied for when proper arrangements shall have been made; but if the removal is insisted upon, the driver must inform the patient or his or her friends, that it will be effected entirely at their risk and cost.

(9) AMBULANCE STATIONS

Rules and Regulations for the Guidance of the Male Staff

The whole of the staff must be up each morning not later than 5.45 A.M.; and in the yard ready for duty at 6 o'clock.

The staff will take their meals at such hours as may be directed by the Superintendent, and approved by the Committee of Management. On no account is any article of food to be taken from the mess-room to the dormitories.

All officers, on every occasion of going on leave, must have a bath, and substitute their own clothing for their uniform before leaving the premises.

All officers on leave are to return punctually to the Station. If any officer is late, the Superintendent may stop his leave on the next occasion, reporting thereon to the Committee at their next meeting.

In the case of an officer being absent without leave, a deduction will be made from his wages, and the matter reported to the Committee.

Any officer, in case of illness, shall apply to the Superintendent of the Station for an order to see the Medical Superintendent of the adjoining Hospital for treatment.

The men's leave and the cleaning of the dormitories will be regulated in detail by the Superintendent.

No man is permitted to sleep in the dormitories during the day.

No smoking is allowed in any other part of the Station but the mess-room.

No naked lights may be carried ; and when matches are used, care must be taken that they are not thrown down whilst alight.

Any officer guilty of misconduct or insubordination will be liable to immediate suspension by the Superintendent, who will report the facts of such suspension to the Clerk forthwith, and to the Committee at their next meeting.

The coachman next for duty must be dressed ready to go out at any time during the day.

When the Superintendent gives the signal every man must be on the alert, and must assist in getting the Ambulance out as speedily as possible.

The coachmen are to use every endeavour to find the patient. When a coachman cannot do so, he should go to the Relieving Officer of the district, or the workhouse of the parish.

At all times the Ambulances must be driven with due care and diligence.

After delivering the patient and nurse the driver must return forthwith to the station without stopping or calling at any place whatever, on pain of instant dismissal.

In case of an accident, should the driver be unable to proceed upon his journey, he must telegraph from the nearest telegraph office the particulars of the accident to the Superintendent.

On no account is a horse to be left alone whilst out with an Ambulance.

(10) Ambulance Stations

Rules and Regulations for the Guidance of the Female Staff

The female staff must be up each morning and at their work at hours to be fixed from time to time by the Superintendent.

The staff will take their meals at such hours as may be fixed by the Superintendent, and approved by the Committee of Management. On no account is any article of food to be taken from the mess-room to the dormitories.

All officers on every occasion of going on leave must have a bath, and substitute their own clothing for their uniform before leaving the premises.

All officers on leave are to return punctually to the Station. If any Officer is late, the Superintendent may stop her leave on the next occasion, reporting thereon to the Committee at their next meeting.

In the case of an officer being absent without leave, a deduction will be made from her wages, and the matter reported to the Committee.

Any officer, in case of illness, shall apply to the Superintendent of the Station for an order to see the Medical Superintendent of the adjoining Hospital for treatment.

The women's leave and the cleaning of the dormitories will be regulated in detail by the housekeeper.

Any officer guilty of misconduct or insubordination will be liable to immediate suspension by the Superintendent, who will report the facts of such suspension to the Clerk forthwith, and to the Committee at their next meeting.

No officer is permitted to sleep in the dormitories during the day.

Officers must retire to their bedrooms by 10.0 P.M., and all lights must be out by 10.15 P.M.

No naked lights may be carried; and, when matches are used, care must be taken that they are not thrown down whilst alight.

(11) METROPOLITAN ASYLUMS BOARD

EASTERN DISTRICT AMBULANCE STATION

Arrangements for working and necessary requirements

When standing in the stables the following is the best time for feeding the horses :—

1st	at	.	.	5.15 (then muck out).
2nd	„	.		7.30
3rd	„	.	.	11.0
4th	„	.	.	3.30
5th	„	.	.	9.0 (Bed down).

Handful of hay in between to keep them amused.

AMBULANCES

Each Ambulance (to be numbered inside, and a corresponding number to be placed over the door or shed) to be fitted with shafts, also splinter bar and pole (splinter bar underside of futchells), grating to each seat, 2 Stanhope lamps for outside (all these lamps to take same sized candles), 1 small lamp inside to hang up, 1 fog lamp to hang up inside also, if necessary (both these lamps are oil lamps).

Stretcher on wheels, with hood (Indiarubber water-beds used for air only), Messrs. Pocock & Brothers, Indiarubber bed, pillow, and seat, ventilator top of Ambulance, outside panels in wood, inside lined with pitch pine (5 cut stuff) and varnished.

Spare beds, seats, and pillows as may be thought necessary, also a few brass fittings for same; 3 inflators; all brass fittings to be universal for the inflators.

Only one door to open at the side on account of stretcher; door to open at back.

One cart very necessary, in case of an accident or break down, to despatch ropes and assistance.

2 setters; 2 shifting spanners, 1 large, 1 small. To be kept in Superintendent's office.

24 spare leather window-blind straps.
24 ,,   ,,   door straps.
1 box of best lamp candles.
Sperm oil for lamps, mixed with 1/3 paraffin (fog and inside lamps).
Lamp cotton, say 25 yards. Scissors for trimming lamps.
Clips and bolts of different sizes, Roller bolts.

1 20-round ladder.
24 bone whistles.
24 ambulance door keys.
One small and one large brand iron, Initials of Station.
1 bottle jack to lift two tons.
1 each axe and saw.
20 Indiarubber mats for inside, 2 ft. + 1 ft. 6 in., Initials of Station.

Harness, etc., say for 12 Horses

6 sets pair horse harness, brass mounted all over. Blinkers and hip straps marked M.A.B.
12 pair pole straps.
12 sets single harness, brass mounted all over. Marked as above.
24 head stalls, 2 hemp reins and 2 logs to each. 24 hemp reins, 12 logs (stock).
12 driver's whips.
12 loin leathers and straps.
12 dandy brushes.
6 each, hard and soft compo. brushes.
6 oil brushes.
6 body ,,
12 water ,,

12 spoke brushes.
6 sets shoe brushes.
12 curry combs.
12 horn mane combs.
12 bath sponges.
24 carriage ,,
24 washing ,,
24 chamois leathers for washing.
24 ,,   ,,   for polishing.
6 stable forks.
6 sieves.
Disinfecting lamps and carbon.
     ,,   powder.
1 tape and 2-foot rule.
Paints, brushes, varnishes, oils, white lead, soft soap, lamp black, silver sand.

2 half-peck measures.

24 whip sockets.

6 doz. of Harris's harness composition.

3 doz. of Harris's harness polishing paste.

6 horse scrapers.

24   ,,   collars, from 21 in. to 22½ in.

6 pairs leather washing leggings, copper riveted.

24 mops and handles.

3 pairs knee pads.

12 horse rugs and rollers.

6 pairs leather collar pads.

6   ,,   ,,   horse boots.

12 ,,   Indiarubber   ,,

3 posting saddles, girths, cruppers, etc.

Smith's tools, iron, steel, timber, boxing machine, washer cutter, vice, anvil, wheel plate, bender, drilling machine, grindstone.

12 split rings to fix on harness for tandem.

Wheelwright's materials.

3 bridles, riding reins, bits, curbs.

3 collars, hames & tugs & hame straps.

3 posting traces with brass coach eyes.

6 sets woollen bands, for horses' legs.

6 pairs galvanised pillar chains.

12 nose bags.

12 wooden pails for stable.

2 doz. galvanised pails, 13 in. × 13 in.

24 bass brooms and handles.

2 clipping combs and scissors.

2 stable barrows.

6 shovels.

6 stable baskets or dung skips.

6 feeding baskets.

Corn bin.

Clark's patent clipping and singeing machine.

50 feet Indiarubber gas tubing.

In harness-room, brackets must be fixed for hanging up harness, saddles, etc., and gas stove for winter.

Recommended that gas lights, in connection with the Station in the stable and yard, be regulated from one tap.

12 pairs clogs with iron tips.   12 pairs driver's gloves (double hand parts).

OFFICE : COMMENCING FROM THE TELEPHONE

Memo.—In all forms of printing see to alteration of the name of Station, etc.

1 cash safe.

1 sq. table, 1 ft. 6 in. × 1 ft. 4 in., with drawers for telephone forms.

1 parchment slate for telephone clerk.

Form 1.—1000 telephone message forms (qy. in black and red).

Form 2.—6 driver's and nurse's note order book (qy. in black and red).

1 large slate, 1 ft. 6 in. long., 1 ft. wide, scratched in to correspond with removal book.

1 particulars of removal book. (No. 62,506.)

Form 3.—700 daily return forms.

  ,,   4.—300 weekly ,,   ,,

  ,,   5.—100 monthly ,,   ,,

  ,,   6.—100 monthly abstract forms.

Form 7.—1000 removal forms to clerks to guardians.

  ,,   8.—100 envelopes (each parish) printed addresses to do.   do.

  ,,   9.—250 tradesmen's account forms (3 pages).

  ,,   10.—500 tradesmens' circulars.

  ,,   11.—Rules and regulations to drivers ⎫ To be framed,
  ,,   12.—Rules and ⎬ glazed, & hung
  directions to nurses ⎭ on the walls of
  the mess room.

  ,,   13—General orders.

500 returns from the opening of station.

1 Stanford's library map of London, in 24 sheets, 6 in. scale, mounted.

1 London Post Office Directory.

1 Suburban Directory.

1 Local Government Directory.

1 book of regulations, and standing orders of the Managers.

1 office desk.

1 ,, slope.

1 office chair.

Stationery of all descriptions necessary, pens, holders; black, blue, red, and slate pencils (Faber's); Indiarubber, 2 rulers, bottle of gum, gum bottle and brush, letter scales, clock for office, 2 basket trays, 2 double ink-stands, 2 waste paper baskets, 6 glass paper weights.

2 books indiced throughout (1 ruled with money columns).

1 wages' book, ruled wide for receipt stamps.

1 petty cash book.

1 Register of officers' book.

1 order check book.

1 receipt book (numbered 1 to 100). No. 62,206.

1 estimate book. No. 61,829.

1 report book.

1 gate book. No. 6189.

1 copying book.

1 ,, press, with damping brush and water well.

1 blotting pad.

Half ream white and red blotting paper.

1 Lett's Diary. No. 44.

1 store's account book. No. 63,764 —15/9/83.

1 inventory book.

1 linen account book. No. 63,763 —14/9/83.

1 day book.

1 summary of day book

1 postage book.

1 stamp for marking linen.

1 brand for marking utensils.

Marking ink.

Lists of clothing forms.

1 large book, indiced for instructions from Norfolk St., paged throughout.

Memo.—All entries at this Station for Smallpox are made in black.

All entries ,, ,, Fever ,, ,, red.

MEN'S MESS-ROOM

If possible the map out of the London Directory should be mounted on rollers, sized and varnished, then hung up in the mess-room.

The rules and regulations should also be framed, glazed, and hung up on the walls of the mess-room.

Coffee	at	7.0 am.,	allowed	10 minutes.		
Breakfast	,,	8.0 ,,	,,	30	,,	
Lunch	,,	10.30 ,,	,,	10	,,	If business permits.
Dinner	,,	12.30 ,,	,,	60	,,	
Tea	,,	4.30 ,,	,,	30	,,	
Supper	,,	8.30 ,,	,,	30	,,	

To bed sharp at 10.15, unless some of the Ambulances are out, when one horsekeeper and the driver next for duty will sit up, the remainder to be compelled to go to bed.

DRIVERS

To commence at six o'clock, start cleaning harness, one appointed in rotation to wash, until some is washed and dried, the others get on with the bridles (bits to be put in the pickle tub); as the harness is washed it is passed on to others to compo., while some clean the brass mountings, etc. ; when finished to be hung up in the harness-room ; one man to be appointed to see the double harness is paired properly and ready for use ; any wanting repairs to be reported to the Superintendent.

See occasionally to all leading and spare harness, or it will get out of order.

When the harness is done and put away, clean gas lamps in yard, harness-room, lavatory, etc., and do such sundry jobs as may be necessary.

At this Station the driver first on turn reports himself ready by 10.30 a.m. to Superintendent ; in all cases, soon as the work is done, the men should clean themselves and be prepared for a call.

Each driver must have a place for his whip, waterproof, knee-apron, etc., so that they can be found when wanted ; when he receives a call he puts his things on the box of the Ambulance, goes to the office for his written orders ; soon as the horse or horses are put to, he will get on the box (in all cases picking up his reins before doing so); when the nurse is inside, she having received her note also, the Superintendent will tell him to proceed on his journey.

Each driver, when engaged, to find one whip. Each driver, in fact all the men, should be provided with some canvas for aprons, to wear while cleaning the harness, etc.

UNIFORMS

2 suits white duck overalls (jacket and trousers only.	1 pair driving gloves, with double hand parts.
1 tweed suit.	1 waterproof knee apron.
1 livery suit (M.A.B.) on collar, metal buttons, coat, sleeve waistcoat, and trousers. *(White cord seams except trousers.)*	1 pair clogs.
	1 bone whistle.
	1 whip. Driver finds one also.
	1 Ambulance key.
1 livery great coat, M.A.B. on collar	1 pair waterproof leggings (dark tweed).
2 caps and oilskin covers (the latter to cover neck).	Weekly, 1 box, Bryant & May's matches, to each of the male staff.
1 waterproof driving coat (white).	Comb and hair brush.

Memo.—Every male engaged should be able to drive if necessary, and at least two of staff should be able to ride horseback.

HORSEKEEPERS

(To get up at 5.15 o'clock, and feed his horses, etc.)

1 worsted cord suit. (Sleeve waistcoat and trousers.)	1 pair driving gloves.
	1 ,, clogs.
1 tweed suit.	1 bone whistle.
2 caps and covers.	1 Ambulance key.
1 livery small coat.	1 waterproof driving coat.
1 livery great coat.	1 ,, knee apron.
1 pair waterprof leggings (dark tweed).	

Washer and Harness Cleaner, same Clothing as Horsekeepers.

MESSMAN AND DRIVER, MALE ATTENDANT AND LAMP CLEANER

Two suits of tweed. Two caps and covers. One livery small coat. And the remainder of the articles mentioned after the livery small coat above.

THE WASHER

To be at work at 5.30. Leather leggings for self and another provided, also clogs. Lamps to be taken off and put in a place appointed ready for cleaning. Commence washing Ambulances that have been used the day previous, under cover if wet or dark, on the outside wash, weather permitting; if in washing he sees anything displaced or broken, to report same to Superintendent. At 6 o'clock when the drivers commence, one, who has been previously told off (in rotation) comes and assists; after all is washed, they must be leathered off; inside sponged and dried, air beds, pillows, and seats blown up, brass work, door and back locks cleaned, transomes oiled, and the Ambulances put back in their proper places, which are supposed to be numbered the same as the Ambulances; see his lamps trimmed and bottom piece locked always before the Ambulance goes out. If any sign of fog, see a fog lamp is placed inside, in the place appointed and fastened; see that splinter bar, pole, and pole straps are in their places behind each single, and the shafts behind each pair, properly numbered same as Ambulance, so that they may be ready at a moment's notice, if orders are given, to change a single to a pair, or the reverse.

The bolts on both singles and pairs for attaching splinter bar or shafts must have the nuts run off occasionally, or they will get set, and, when wanted to change, be difficult to move.

When the Ambulance returns from a journey, to wash the same decently; at any rate, to sponge the panels, and leather off; see to inside—that bed, pillow, and seat is blown up, and everything made decent for use again that day if needed.

To be prepared (when drivers are all out) to go with an Ambulance, if ordered to do so by Superintendent.

NURSES

The nurses' breakfast at	.	.	.	8.0
Dinner at	.	.	.	12.30
Tea at	.	.		4.30
Supper at	.	.	.	8.30

And, unless out in an Ambulance, to bed at 10.15 sharp.

The nurse first on turn to be ready when she comes down to breakfast; all should be ready by 9.30, having previously made their beds, emptied slops, dusted room; it being understood they keep their own sleeping-room clean, except scrub the floor and clean the windows.

The housemaid brings the nurses' meals, and puts them on a table outside their mess-room door, the nurses take it from there; after they have had their dinner, the things are placed on the table outside by the nurses for the maid to take away.

The Superintendent's wife, as housekeeper, will find ample employment in seeing the house is kept clean, cooking done properly, that no waste occurs, that the nurses do not go to their dormitories to sleep in the day-time, take charge of the linen, mark and repair same, take account of and send soiled linen to the laundry, checking and seeing to the airing of same on its return, and the general duties appertaining to a housekeeper.

SUPERINTENDENT

Memos.

To study the map, principally the district he will have to work, to be enabled to instruct the drivers where they have to pick up.

To make himself acquainted with the names of the Relieving Officers, so that by reference to the Local Government Directory he will see from what parish the patient comes, in case the information is not sent him by telephone.

On receipt of telephone message, the Superintendent blows on his whistle twice distinctly, that being the signal for the head horse-keeper, the Superintendent instructs him thus :—

FEVER
{ F. a pair } That is, a pair of horses and a Fever Ambulance.
{ F. single } ,, Fever Ambulance, with one horse.

or

SMALLPOX
{ Small, a pair } That is, a pair of horses and a Smallpox Ambulance.
{ Small, single } ,, Smallpox Ambulance, with one horse.

Everyone at the Station being on the alert when the Superintendent whistles twice :—The washer brings out an Ambulance to the place appointed for putting the horses to ; the horsekeepers put the harness on the horses, everyone assisting to do what they can (if at night-time, one lights the outside lamps, another the inside one, etc.) ; no excuse from this, no waiting to finish dinner, the Ambulance is to be sent out with all due speed. Three calls on the whistle will have warned the nurse, who is supposed to be ready waiting.

While the horses are being put to, the Superintendent makes out the driver's and nurse's notes, on which he has written the particulars received by telephone.

The drivers are to understand they must use every endeavour to find the patient ; sometimes the number may be wrong, or the street not spelt correctly ; if he cannot do so, he should go to the Relieving Officer of the district, or the workhouse of the parish ; he should perfectly understand, if he came back to the Station without removing the case, another horse would be put to at once and he would be sent out again, unless the Superintendent thought other-wise. Every effort should be made to get to the patient as soon as possible, and deliver the case at the Hospital with care.

The men's leave and cleaning the dormitories would be for the Superintendent to consider in detail.

It should be a strict rule that neither men nor nurses be permitted to go to their dormitories to sleep during the day.

It is indispensable the whole of the male staff be able to read and write legibly.

No smoking permitted while at work.

All clothing and utensils, before being served out, to be branded with initials of station.

CHAPTER IX

PRIVATE associations have been found of much service in assisting Local Authorities and their officials in carrying out the measures necessary to prevent the spread of infectious disease. As long as the behaviour of infectious diseases is imperfectly understood, or the importance of taking measures to prevent them is not fully appreciated, so long will such associations be found useful. At present they are of importance in bringing the matter before the public. But when the importance of preventing disease is properly understood and appreciated, private associations for this purpose will not be required any more than such assistance is now necessary for the suppression of crime, or for the collection of the Inland Revenue.

Mrs. Francis Johnstone,[1] the manager of the Sanitary Aid Association at Hastings, stated in her evidence before the Smallpox and Fever Hospitals Commission, that "the function of that Association is to supplement the resources of the medical officers of

[1] Fever and Smallpox Hospitals Commission, Minutes of Evidence, pp. 209, 210.

health. The medical officers of health can give orders, but they cannot relieve, and there is no woman teacher ; they can give orders, but not detailed instructions. The inspectors are supposed to give instructions, and to a certain extent they do. Sometimes, indeed, the inspectors are very vigilant and good, but as a rule you find that the man has given orders in general terms, which, until our woman teacher has explained by giving detailed instructions, are not thoroughly understood." The Association usually gets information from district visitors in regard to houses where infectious disease prevailed. The clergy, with their whole staff of teachers, paid and unpaid, and their lay helpers, are the great agency for information. When the teacher of the Association visits the mother of a sick child, if the case is to remain at home, she "teaches the mother the practice of disinfection at every point, so that from the moment of the child being isolated in a separate room, nothing that has not been disinfected leaves the room. The teacher continues to visit the case during a period of eight weeks ; no matter how soon the child may recover, so far as being well, during eight weeks our care continues." "If the case is removed to the hospital we ensure disinfection. As I have said, frequently it is all done rightly without us, but sometimes it is not. We complete those performances as they may need completion, and if the child comes out of the hospital sooner than we can venture to let it go back to school, we attend to that."

In the case of the poor certain relief is given if the orders of the medical officer of health are really carried out. " The motive power is the conditional relief; the relief that might be given, if no respect is paid to the fulfilment of instructions, is stopped. That is why we take it off the Church alms' fund; the clergy support us, and we take all those cases entirely off their hands."

The average expenditure for the protection of all the people [over 45,000] is £100 a year. We have only one salaried person in the society, and that is the teacher, who has a salary of £40 a year. The remaining £60 does for relief of all kinds, printing, and disinfectants sometimes, as well as other expenses.

This Association, according to Mrs. Johnstone, has been successful on many occasions in preventing cases of infectious disease from spreading, and of great service in supplementing the resources of the medical officers of health. The following are some of the rules of the Association :—

RULES OF THE SANITARY AID ASSOCIATION AT HASTINGS [1]

For Boys' Schools

Scarlet Fever

It is quite possible to check scarlet fever if proper means be taken to destroy the emanations of the sick, so that they shall not infect the healthy. For this purpose the following recommendations have been compiled from the best authorities on the subject, by the

---

[1] Appendix G to Smallpox and Fever Hospitals Commission, Minutes of Evidence, pp. 339, 340, 341.

direction of the Committee of the Local Board of Health for the
district of the city and county of Bristol, and, with a few small but
very important additions or modifications, are adopted by the
Hastings and St. Leonards Sanitary Aid Association :—

1. If a case of scarlatina, or scarlet fever, or bad sore throat,
appear in your house, apply immediately to a medical man, and
send information of it to the medical officer of health, and to the
sanitary aid manager.

2. If possible, separate the patient immediately from the rest of
the inmates. A room at the top of the house is, as a rule, the best
sick-room.

3. Let the room in which the patient lies be stripped of all
carpets and curtains.

4. Let all the discharges, of whatever kind, be received on their
very issue from the body into a disinfectant, such as Calvert's
powder, chloride of lime, carbolic acid, or Condy's fluid, and continue this from the first discovery, or even suspicion, of scarlet
fever until eight weeks from that date, no matter how much sooner
the patient may appear perfectly well, and his skin quite free from
any remainder of peeling.

5. Let small pieces of rag be used instead of pocket handkerchiefs for wiping the mouth and nose ; each piece after being once
used should be immediately burnt.

6. About the fourth day of the eruption let the surface of the
body be well rubbed with camphorated oil daily, the oiling to be
continued until the patient is able to take a warm bath, in which
the whole skin should be well scrubbed with disinfecting carbolic
acid soap.

N.B.—In no case must the sanitary aid visitor order oiling or
bath without consulting the medical attendant of the case ; with his
permission the bath, including washing the head, should be used
twice or thrice a week to the end of the eighth week.

7. The patient may, if a clean case, in clean clothes re-enter the
family after five weeks from first appearance of rash, but he must
not kiss any one, and must still sleep apart.

8. A large vessel, containing Condy's fluid in the proportion of
one ounce to every gallon of water, should be kept in the room.
All bed and body linen, on its removal from the person of the
patient, to be immediately placed therein.

9. In case of death, the corpse should be smothered with carbolic powder, and speedily buried.

10. No child having had the scarlet fever should be allowed to re-enter a school without a certificate from the medical officer of health stating that he can do so without risk to others, and this certificate must not be asked for until the eighth week is past, and rule 11 completely carried out.

11. On the recovery or removal of a patient, all floors, walls, and ceilings should be fumigated, scraped, and cleaned. For fumigating infected rooms and their contents, nothing is better than sulphur. A quarter of a pound of brimstone, broken into small pieces, should be put into an iron dish (or the lid of an iron saucepan turned upside down), supported by a pair of tongs over a bucket of water. The chimney and other openings are then closed with paper pasted on, and a shovelful of live coals is put on the brimstone. The door is then quickly shut, the crevices covered with paper and paste, and the room kept closed for five or six hours. After this a thorough cleansing should be effected, everything washable should be washed, and all other things be cleansed by proper means.

N.B.—There is an important distinction to be observed between a clean and a foul case of any fever. A clean case is one unattended by any description of involuntary discharge. A foul case is the reverse of this. In a foul case no half quarantine, such as is indicated in Rule 7, can possibly be allowed ; for instance, an infant too young to exercise self-control as to the ordinary discharges, must remain in strict isolation during eight full weeks, and any subject thus affected in respect of any discharge whatever, must be similarly treated while so affected.

The before mentioned rules apply to cases not constitutionally diseased. A case with running sores of any kind must be entirely isolated twelve weeks, or there can be no security against the spread of the disease.

Daily fumigation of the sick-room, and of the house in which it exists, may very beneficially be performed with sticks with the bark on, sold for lighting fires, or with chips dipped in tar, when live sticks cannot be procured.

A person nursing a fever case, who has never had the fever, should never swallow while changing soiled clothing or attending to discharges, but should cleanse nose and mouth and throat, and presently gargle with water made bright pink with Condy's (red cross) remedial fluid.

" Is to be dealt with in all respects like scarlet fever, with the substitution for camphorated oil of a skin dressing of charcoal powder and olive oil, mixed to the consistency of paint, and applied with a brush. Spots kept covered with this will not pit. No. 4 rule must be applied in smallpox cases not less than six weeks from first discovery."

Typhoid Fever

" This disease, which is of an infectious nature, is easily prevented from spreading if proper means be taken to remove the original source of infection, to isolate the patients, and to destroy all emanations from their persons. For this purpose the following recommendations have been compiled from the best authorities on the subject, by direction of the rural sanitary authority of the Clifton Union, and are adopted without any change by the Sanitary Aid Association for the Borough of Hastings.

1. If a case of enteric, typhoid, gastric, or low fever (different names applied to the same disease) appear in your house, send immediate information thereof to the Medical Officer of Health of the district.

2. Have your house inspected by a competent person, and make sure that no sewer or drain gas can enter into any part of it. Any defect of this may render you liable to infection from a case of fever at a distance from your residence.

3. Have your water supply for domestic use examined, and if in the slightest degree contaminated with sewage matter, immediately discontinue the use of it. Rain water received in cisterns or barrels above ground, and filtered through a common charcoal filter, is always safe. Water contaminated with sewage derived from an infected source is one of the most common causes of the disease.

4. Let the patient be isolated in a well ventilated room, without carpets or curtains, and, if possible, at the top of the house.

5. Let all discharges from the patient, especially those from the bowels, be received into a disinfectant; the most convenient are chloralum or Calvert's powder. A piece of gutta-percha sheeting or oilcloth should be placed under the blanket, under the breach of the patient, to prevent the discharges soaking into the bed.

6. The bed and body linen, and all other infected clothing, should be plunged in water containing four ounces of Calvert's carbolic acid (No. 5) to every gallon of water, and afterwards boiled before being washed.

7. As soon after recovery as the patient is able to bear it, he should take a tepid bath, or be washed with warm water and carbolic acid soap; he may then re-enter the family with safety.

8. Attendants on the sick should be scrupulously clean, and frequently wash their hands with a disinfectant; and they should carefully abstain from touching

any article used for the food of man, such as milk, etc. Their personal clothing should be treated as infected articles.

9. Any article of food which has been exposed to infection in the patient's room should, when not consumed by the patient, be destroyed.

10. Every closet in the house, and every eject leading into a drain, should be disinfected twice daily by throwing into it a handful of green copperas. As the germs of this disease are most generally disseminated by means of the drains, every system of drains receiving the evacuations of a typhoid patient should be kept constantly charged with this inexpensive chemical.

11. As some persons from peculiarity of constitution take this disease in an extremely mild form, hardly recognisable as fever, all persons residing in a house containing a typhoid patient, who are suffering from the slightest indisposition, and especially if it is attended with diarrhœa, should confine themselves to the house, treat their own evacuations as infected, and scrupulously abstain from using their neighbour's closets. Persons of this class are often the means of spreading this and other diseases in public factories, where closets are used in common. All such closets should be at all times flushed and disinfected twice daily.

12. Good ventilation is the best disinfectant of the air of the sick-room.

13. On the recovery or removal of a patient, all

floors, walls, and ceilings should be fumigated, scraped, and cleaned. For fumigating infected rooms and their contents, nothing is better than sulphur. A quarter of a pound of brimstone, broken into small pieces, should be put into an iron dish (or the lid of an iron saucepan, turned upside down), supported by a pair of tongs over a bucket of water. The chimney and other openings are then closed with paper pasted on, and a shovelful of live coals is put upon the brimstone. The door is then quickly shut, the crevices covered with paper and paste, and the room kept closed for five or six hours. After this a thorough cleansing should be effected ; everything washable should be washed, and all other things be cleansed by proper means.

14. Any further advice of a public character, required for carrying out the above or other sanitary precautions, will be given on application to the Medical Officer of Health or Inspector of Nuisances of the district."

Penalties

" 1. Catching or infectious diseases are scarlet fever, smallpox, typhus, typhoid, and relapsing fevers, and measles [diphtheria and whooping cough].

2. By the Sanitary Act of 1866, a penalty of £5 is inflicted on persons who wilfully or negligently are the means of spreading infectious and contagious diseases among their friends or neighbours.

3. It is illegal to use any public cab for the conveyance of a patient to a hospital or anywhere else

without telling the driver that it is a case of infectious disease.

4. The driver of a cab may refuse to take any such person unless he is paid a sum of money sufficient to defray the expenses of disinfecting his cab.

5. Any cabman taking another fare after conveying an infected person, without previously disinfecting his cab, is liable to a penalty of £5.

6. It is illegal for an infected person to go, or for any person to take or send anyone suffering from an infectious disease, to any public place, such as the waiting-room of a hospital or a dispensary, or to a school, or a church, or a chapel, or a theatre, or omnibus or other public carriage, so as to endanger any other persons, whether adults or children.

7. It is also illegal for any person to give, lend, sell, or move to another place, or expose any bedding, clothing, rags, or other things which may have become infected and are liable to convey any contagious diseases to another person, unless such things have been previously disinfected.

8. It is also illegal to let any house, room, or part of a house, in which any person has been ill with any infectious disease, until it and all articles in it have been properly disinfected. The same law also applies to public-houses, hotels, and lodging-houses. The penalty for disobedience in these cases is £20.

9. The means of conveying persons to the fever hospital or smallpox hospital can be ascertained by applying to the Medical Officer of Health."

APPENDIX

HOSPITAL PLANS

"THIS hospital was erected in 1889 by the Warwick Joint Hospital Board, the constituent authorities of the district being the urban districts of Leamington Spa, Warwick, Kenilworth, Lillington, and Milverton, and the rural sanitary district of Warwick Union. Since the formation of the joint district, the districts of Lillington and Milverton have been amalgamated with the borough of Leamington.

The site is six acres in extent, two and a half acres of which are enclosed and appropriated to the hospital use. It stands on high ground in the middle of open fields, and is about midway between the towns of Warwick and Leamington.

The population within the joint district was, when the hospital was erected, about 55,000, and of this number some 44,000 were within a radius of three miles from the hospital site.

The area of the district is 44,507 acres, and its ratable value £340,000. The accommodation provided is equal to one in every 2000 of population.

The buildings are four in number, and each is entirely detached from the remainder. They are—

1. The administration block;
2. The isolation block;
3. The ward block; and
4. The laundry block.

The site is enclosed on three sides by a high wooden fence; the front towards the high road having a dwarf wall and an iron railing.

The two blocks occupied by patients and the laundry block

---

[1] Read in the Architectural Section of the Seventh International Congress of Hygiene and Demography, held in London, August 1891, by Keith D. Young, F.R.I.B.A.

are each 40 feet distant from the boundary, and an inner fence at a distance of 40 feet from the front gates prevents patients from approaching too near the latter.

The administration block is the only one of the four buildings which is two storys in height, the rest being one story only. It contains on the ground floor a sitting-room for the matron and nurses; a room, with a lavatory and w.c. attached, for medical officers; linen room, w.c. for nurses, kitchen, scullery, and larder, with wood and coal stores, and servants' w.c. in the yard. At one corner of the kitchen is a serving hatch, opening into a covered porch, at which the meals for the patients are given out.

On the upper floor are bedrooms for the matron, nurses, and servants, and a bath-room.

The *Isolation Block* is divided into two equal parts by a wall, and the arrangements on one side of the wall are an exact counterpart of those on the other side, the entrance to one side being on the east, and that to the other side on the west.

Each half of the building, therefore, contains a large ward for three beds, two small wards for one bed each, a nurses' duty-room, and a w.c. and slop-sink. These rooms communicate with each other by way of an open verandah, roofed over at the top, but quite open in front.

Each large ward is 36 feet long by 18 feet wide, the smaller wards being 18 feet by 12 feet, and all are 12 feet high. The allowance of floor space is 216 feet per bed, and of cubic space 2592 feet per bed. The large wards are lighted by three windows at each end, the smaller ones by one window on each side, being in the proportion of one square foot of window surface to about 65 feet of cubic space. The windows, which form the principal means of ventilation, are divided into two parts by a transom which is fixed about 1 foot 6 inches down from the head of the frame. Below the transom are ordinary double hung sashes provided with a deep bottom rail and a cill board, which permits of the lower sash being raised and a current of air admitted in a vertical direction between the two sashes, at the same time preventing free access of air at the cill level. Above the transom is a 'hopper light,' hung on hinges at the bottom to fall inwards, and provided with glazed cheeks at the sides to prevent down draughts.

In addition to the windows, openings are made at the floor-level behind each bed, and provided with Ellison's radiator ventilators; there is also an extraction flue in each ward carried up alongside the smoke flue, from which it is separated by iron plates. The inlet to the flue is at the ceiling level, and a Bunsen burner is provided with a view to produce an upward current when the fire is first lighted.

The wards are each provided with a Boyd's hygiastic ventilaing grate; these grates have at the back of the fire an air chamber which is supplied with fresh air from the outside. The air, being warmed by contact with the heated fire-brick back of the stove, passes into the room through a grating above the fireplace.

The walls of the ward are lined to a height of 5 feet with tinted glazed bricks, above which they are plastered and distempered. The floors are laid with yellow deal in 3 feet widths, ploughed and tongued.

The vertical angles of the walls, the horizontal angles at the junction of floors and walls, and of walls and ceilings, are all rounded, so also are all the angles of door panels and of the windows, and in the finishing of the doors and windows rounded fillets only are used, no recessed mouldings being used anywhere.

The nurses' duty-room is provided with a small range with oven and boiler, and hot water is laid on from the latter to the sink, the movable bath, in corridor, and the slop-sink. There is also in this room a small dresser and a glazed porcelain sink.

Outside the duty-room is a recess, where the movable bath stands. A glazed fireclay sink let into the floor takes the waste, and taps fixed to the wall afford the supply of hot and cold water.

The water-closet and slop-sink are placed in projecting buildings, entered from the verandah. The walls of these offices are lined with glazed bricks, and the floors are of cement. The slop-sinks are of porcelain, provided with a flushing rim, in addition to the hot and cold water taps. The closets are Hellyer's pedestal hygienic, the trap and basin being made in one piece, of porcelain, and are fitted with 3-gallon flushing cisterns, and hard wood rim seats hinged at the back.

Ward Block.—This building is entered from the centre, and affords accommodation for twelve patients, all of one disease, the beds being equally divided between the two sexes. In the open porch at the entrance are two doors; one of these gives access to the entrance lobby, the other being an outer door to the bathroom. The object of the latter is to enable a patient on being discharged, to leave the building directly from the bath-room. The bath-room thus becomes a discharging-room ; not perhaps an ideally perfect arrangement, but certainly a better one than if the patient had to re-enter the ward after his final bath.

To the left of the entrance is a small cupboard with a window for food, and opposite are cupboards for patients' ward clothes and linen. Between the wards is the nurses' duty-room, in which is a small range with boiler for supplying hot water to the bath, sinks, and lavatory basins, a dresser, and a porcelain sink.

The wards are each of them 36 feet long and 26 feet wide, and contain six beds each. To each bed is allotted a floor space of 156 feet, and a cubic space of 2028 feet, and the distance from centre to centre of each bed is 12 feet. The window area is in the proportion of about 1 foot of window to every 60 feet of cubic space. The means of ventilation are similar to those adopted in the isolation block. The grates also are similar, but in these wards they are placed in pairs, back to back, in the centre of the floor, with descending flues carried under the floor to vertical chimneys in the outer wall. The latter are swept from the outside.

The water-closets and slop-sinks are placed in the projecting buildings at the end of the wards, from which they are separated by cross-ventilated lobbies.

The construction of these wards, as regards internal finishing of floors and wall surfaces, etc., is in all respects similar to that of the isolation block.

A speaking-tube connects the duty-room with the administration block.

The laundry consist of a wash-house fitted with the usual appliances for washing, an ironing-room, and a drying-room heated by the flue of the ironing stove, and fitted with a radial drying horse. Adjoining is a w.c. and a coal store.

The disinfecting-house is divided into two parts by a brick wall. The apparatus, which is one of Washington Lyon's high-pressure steam machines, projects on each side of this wall, so that the infected clothes are put in one chamber, and when disinfected are taken out by the door in the other chamber.

The mortuary is a plain sky-lighted room, arranged for use when necessary, as a post-mortem-room. The ambulance house affords accommodation for a one-horse ambulance.

Drainage and Water Supply.—The drainage system is a dual one, the rain water being separated from the sewage, and stored in a tank for use. The drains are all laid with glazed stoneware pipes jointed with cement, with manholes at each junction and change of direction. Each length of pipe between the manholes was separately tested with water before being covered up, and all the pipes are laid on and partly embedded in concrete. At the end of every length of drain is a Doulton's 30-gallon automatic flushing tank, and at the outfall to the public sewer is a large flushing tank fixed by the town authorities. These flushing tanks were specially necessary in this case, as not only is the quantity of sewage discharged at one time necessarily small, but the sewer conveying the hospital sewage to the town sewers has to traverse a long distance before it comes near any other buildings. The soil pipes are carried up above the eaves of roofs, retaining their full diameter, as ventilating shafts.

The rain water from the roofs is all collected into an underground tank, first passing through a filter chamber formed partly of coarse and fine gravel, and partly of charcoal. From the tank it is pumped for use in the scullery of the administration block and the wash-house.

The cost of the hospital was as follows :—

		£	s.	d.
1.	Land	900	0	0
2.	Buildings, including laundry fittings, disinfecting apparatus, roads, paths, fences, drains, and professional charges	7635	10	6
3.	Gas-mains, from nearest point up to site	84	4	4
4.	Water-mains	161	10	11
5.	Sewer	299	4	9
6.	Furniture	293	17	7
		£9374	8	1

To meet this outlay, loans were obtained from the Loans Commissioners to the extent of £9316 at $3\frac{1}{2}$ per cent, repayable in thirty years. The interest and repayment of principal amounts to an annual sum of £630, which is met by a rate equal to seven-sixteenths of a penny in the pound.

As regards cost of maintenance, it is not very easy to arrive at a definite figure in regard to a hospital which is liable to be occupied or not in so absolutely uncertain a way as this; but assuming the wards were empty for twelve months continuously, it is estimated that the cost of maintenance for that period would amount to £370. This sum includes the following expenses :— Clerk, medical officer's retaining fee, steward, matron, caretaker, outdoor porter, nurse, rations for residents, coal, gas, and water, rates and taxes. For the statistics as to cost and for much valuable information, I am indebted to Mr. Alderman Wackrill, the Chairman of the Joint Board."

For permission to publish the above information, for the ground plans, and for photographs of the hospital, I am indebted to Mr. Keith D. Young, the Architect of the hospital, and to Mr. Alderman Wackrill, the Chairman of the Joint Board.

ADMINISTRATIVE BLOCK.

WARD BLOCK.

HEATHCOTE INFECTIOUS HOSPITAL, LEAMINGTON.

Isolation Block.

HEATHCOTE INFECTIOUS HOSPITAL, LEAMINGTON.

GROUND FLOOR PLAN.

THIS hospital was opened in 1877, since which date, however, the amount of accommodation available has been materially increased.

The site consists of a somewhat irregular oblong, having throughout a width of 195 feet, and an average length of 360 feet; in all, somewhat over 1½ acres. It is situated somewhat over half a mile from the centre of the borough. The soil is sand, some 12 feet deep, overlying clay.

The present buildings, together with that part of the recreation ground enclosed by them, cover about two-thirds of the site, leaving a space at the northern extremity for such further extension as may be found necessary. They consist of:—

1. An administrative block.

2. Two ordinary ward pavilions, which are connected with the administrative block by means of a covered passage open at the sides.

3. A special pavilion containing two wards.

4. A lodge, which is built close to the hospital entrance.

5. Two groups of buildings: one containing a laundry, disinfecting chamber, ambulance shed, mortuary, and store for wood, coals, etc.; the other containing two van sheds, and a store for garden tools.

The administrative block is a substantial two-storied building. This building, as well as the ward pavilions, is built of brick with stone facing, resting on a bed of concrete. The external walls are 16 inches thick, including a 2 inch cavity. In the walls of the wards this cavity intervenes between an outer layer of 9 inches, and an inner one of 5 inches. The hospital contains 28 beds in all, and it affords accommodation for the simultaneous treatment of patients of both sexes, suffering from three infectious fevers.

The population in 1881 was 41,456, and the ratable value was £129,673.

The expense incurred in the construction and furnishing of the hospital was as follows :—

	£	s.	d.
Purchase of land for site . .	£300	0	0
Release of right in Aiken Street .	50	0	0
Erection of 1st portion of Hospital	2182	5	1
„ 2nd „	3320	10	0
Hot water apparatus .	79	18	0
Entrance gates, etc. . .	95	0	0
Kitchen range, grates, chimney pieces	120	5	9
Gasfittings .	80	18	0
Water-fittings . .	27	8	6
Furniture, bedding, etc.	105	9	7
„ „ plumbing, painting, etc., in 1878-80 .	355	3	3
	£6716	18	2
Disinfecting stove	125	0	0
Ambulance	74	0	0
	£6915	18	2

The above particulars, as well as the plans, were copied by permission of Dr. Thorne Thorne from supplement to the Tenth Report of Medical Officer of Local Government Board, 1880-81. Re-issued 1893, where further information may be obtained.

BOROUGH OF WARRINGTON

INFECTIOUS DISEASES HOSPITAL.

AIKIN STREET

SCALE

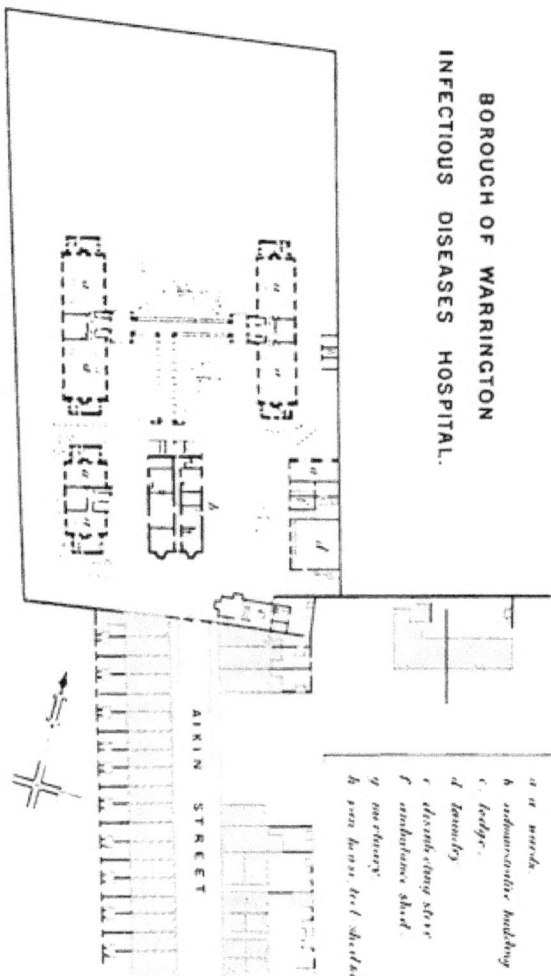

a *wards.*
b *administrative building.*
c *lodge.*
d *laundry.*
e *disinfecting store*
f *ambulance shed*
g *mortuary*
h *convalescent shelter*

SIX BED WARD

SIX BED WARD

NURSES DAY ROOM

BED

EARTHENWARE STORE

ENTRANCE

STORE & SINK

BATH ROOM

SINK

W.C.

MOVEABLE BATH

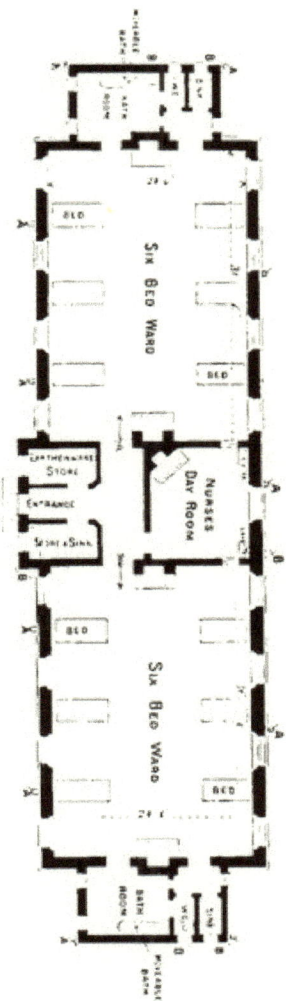

PLAN OF A TWELVE BED PAVILION

Scale 1/8 Feet to an Inch.

A. Down Spouts discharging in trapped Gullies
B. Rain Pipes 4"

The Longton
Borough Surveyor

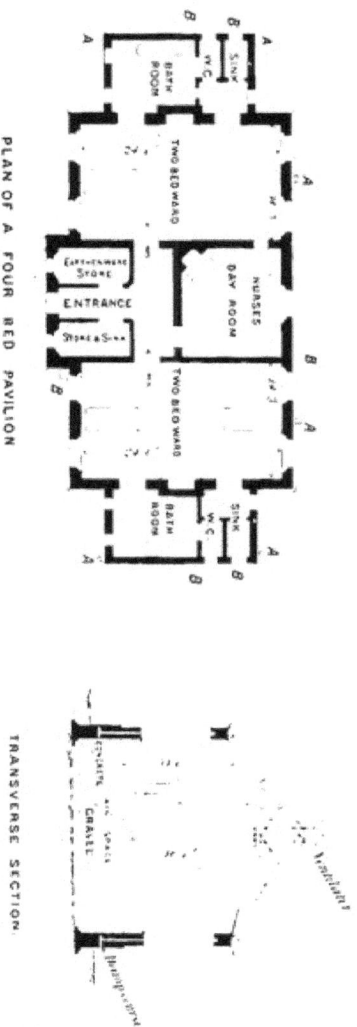

BOROUGH OF WARRINGTON
INFECTIOUS DISEASES HOSPITAL.

PLAN OF A FOUR BED PAVILION

TRANSVERSE SECTION

SCALE 16 FEET TO AN INCH.

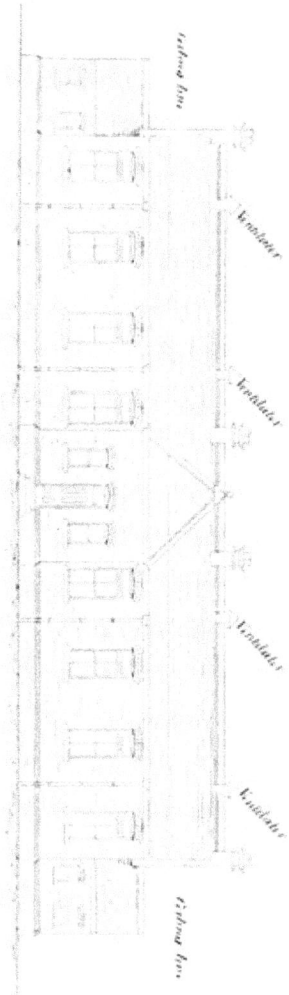

BOROUGH OF WARRINGTON
INFECTIOUS DISEASES HOSPITAL.

ELEVATION OF A TWELVE BED PAVILION.

SCALE 16 FEET TO AN INCH

BOROUGH OF WARRINGTON

INFECTIOUS DISEASES HOSPITAL.

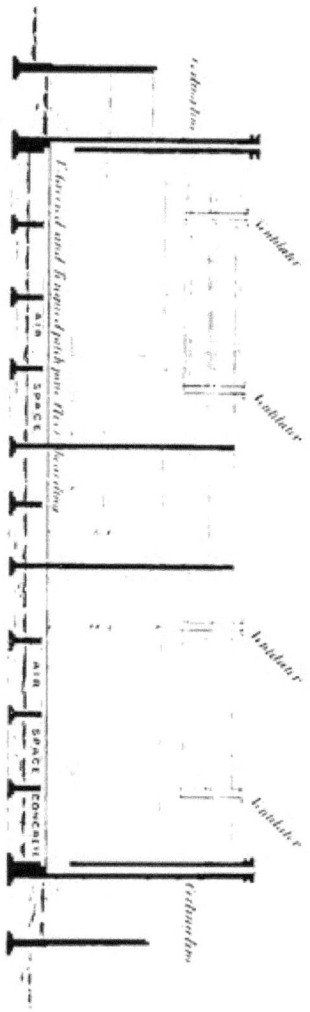

LONGITUDINAL SECTION OF A TWELVE BED PAVILION

SCALE 16 FEET TO AN INCH

AIR SPACE

AIR SPACE CONCRETE

EALING LOCAL BOARD ISOLATION
HOSPITAL

Q

THIS hospital was erected in 1885. The site is one and a half acres in extent, and is surrounded by a high wall. It is beautifully laid out, and planted with trees, shrubs, and flowers. The area of the site is sufficient for an additional block of wards if that be found necessary. Part of the site is covered with concrete for the erection of a tent or other temporary building.

The administrative building is sufficient to meet the requirements of future extension.

The population of the district is 25,000, and the ratable value £168,000.

The number of beds in the hospital is 18.

The cost of building the ward block, administration block, porter's lodge, laundry, disinfecting chamber, ambulance shed, mortuary, gas mains, sewers, and walling of the entire site, was £4750. The cost of the furniture was about £500.

I am much indebted to Mr. Charles Jones, M.Inst.C.E., F.S.I., Ealing, for the plans and particulars of this hospital, as well as for the photographs from which the engravings were taken.

ISOLATION HOSPITAL, EALING.

WARD BLOCK.

ADMINISTRATIVE COTTAGE.

ISOLATION HOSPITAL, EALING.

LAUNDRY AND WASH-HOUSE. AMBULANCE SHED.
DISINFECTING CHAMBER. MORTUARY.

WARD BLOCK.

EALING LOCAL BOARD

ISOLATION HOSPITAL,

— 1885. —

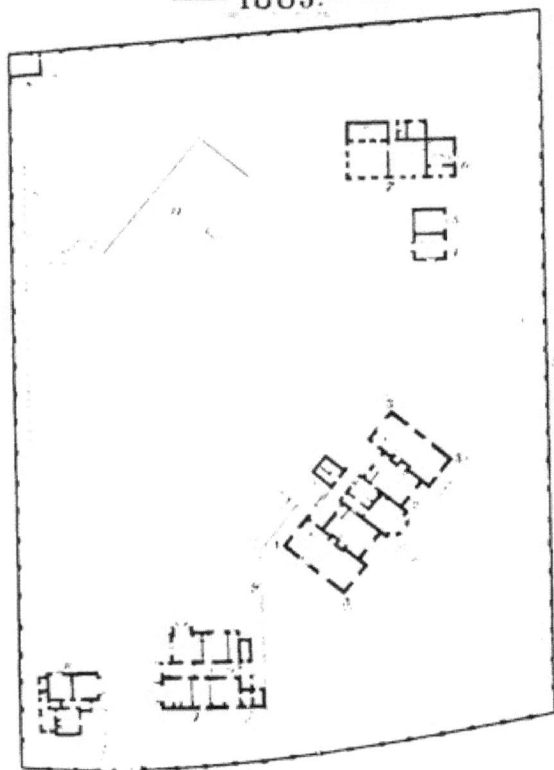

EXPLANATION

No 1 Administration Block
2 Hospital Block
3 Area under ditto to feet
4 Mortuary   Wardings
5 Ambulance
6 Disinfecting Chamber
7 Laundry
8 Conservatory
9 Covered Corridor
10 Porters Lodge
11 Generals Block for tents

SCALE OF FEET.

C. Jones C.E.
Surveyor.

ISOLATION HOSPITAL, EALING.

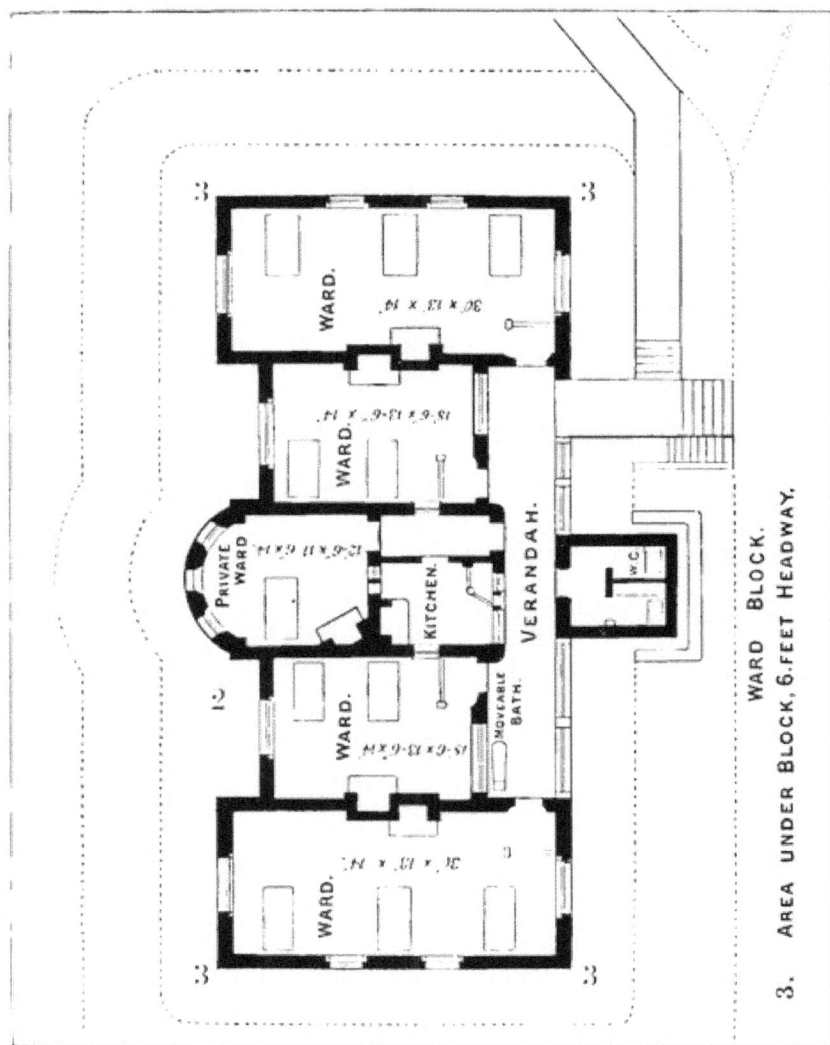

WARD.

30' × 13' × 14'.

WARD.

18'·6" × 13'·6" × 14'.

PRIVATE WARD.

10'·6" × 11' × 6" × 14'

KITCHEN.

VERANDAH.

MOVEABLE BATH.

W.C.

WARD.

18'·6" × 13'·6" × 14'

WARD.

30' × 13' × 14'.

WARD BLOCK.

3. AREA UNDER BLOCK, 6-FEET HEADWAY.

STONEHOUSE ISOLATION HOSPITAL, LANARKSHIRE

This hospital is in course of erection. It is one of four hospitals to be provided in the middle ward of Lanarkshire.

The area of the site is 3 acres, and the feu-duty £12 per acre. The population intended to be served by it is about 10,000, and the total estimated cost of the hospital is £5721.

As may be seen from the ground plan, the hospital consists of— (1) An administrative block; (2) Laundry and disinfecting block; and (3) Two ward pavilions; all detached and at safe distances apart. Ample space is left for future extension.

The administrative building has accommodation in excess of the present requirements, thus providing for further enlargement.

The isolation-wards pavilion, is, with slight alterations, in accordance with the design recommended by the Local Government Board. A verandah is not provided, bringing this part of the building perhaps more in keeping with the climate of Scotland.

In connection with and at the entrance to the fever-wards pavilion, rooms are provided for the disinfection of patients before being discharged from the hospital. These consist of an undressing-room, a bath-room, and a dressing-room. An additional w.c., a store-room, and coal cellar are also provided.

For plans and particulars of this hospital I am indebted to Dr. M'Lintock, Medical Member of the Local Government Board for Scotland.

STONEHOUSE ISOLATION HOSPITAL, LANARKSHIRE.

GROUND PLAN.

STONEHOUSE ISOLATION HOSPITAL, LANARKSHIRE.

GROUND PLAN. ADMINISTRATION.

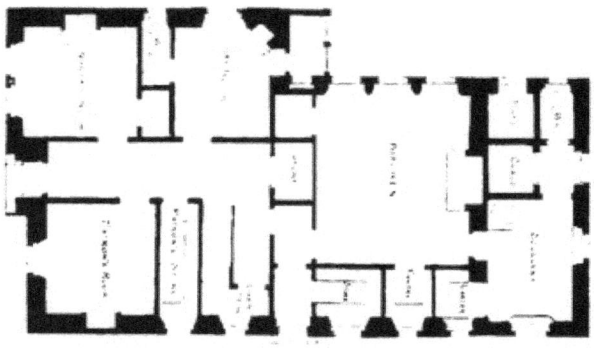

FIRST FLOOR. ADMINISTRATION.

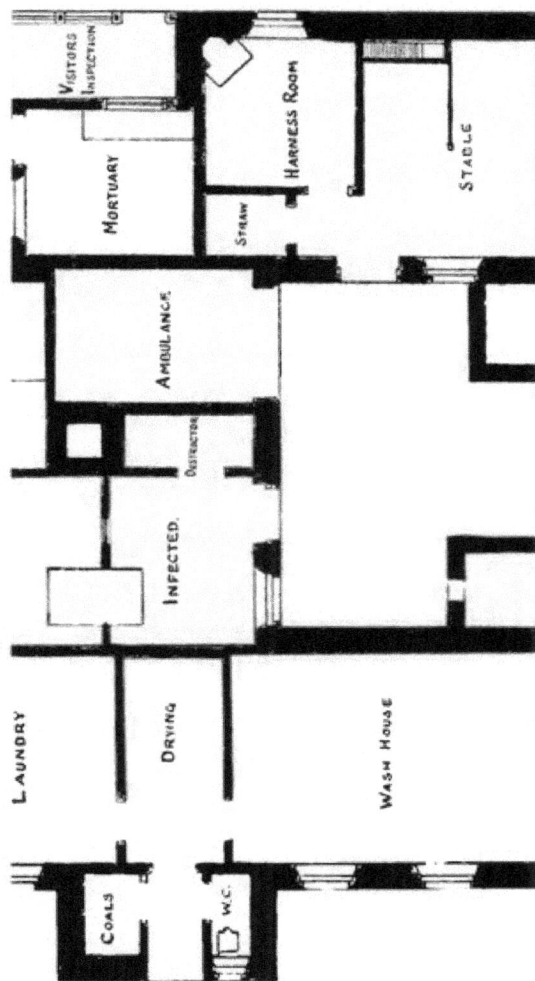

SE ISOLATION HOSPITAL, LANARKSHIRE.

VISITORS INSPECTION

MORTUARY

HARNESS ROOM

STABLE

STRAW

AMBULANCE

DISINFECTOR

INFECTED.

LAUNDRY

DRYING

WASH HOUSE

COALS

W.C.

LAUNDRY BLOCK.

STONEHOUSE ISOLATION HOSPITAL, LANARKSHIRE.

ISOLATION PAVILION.

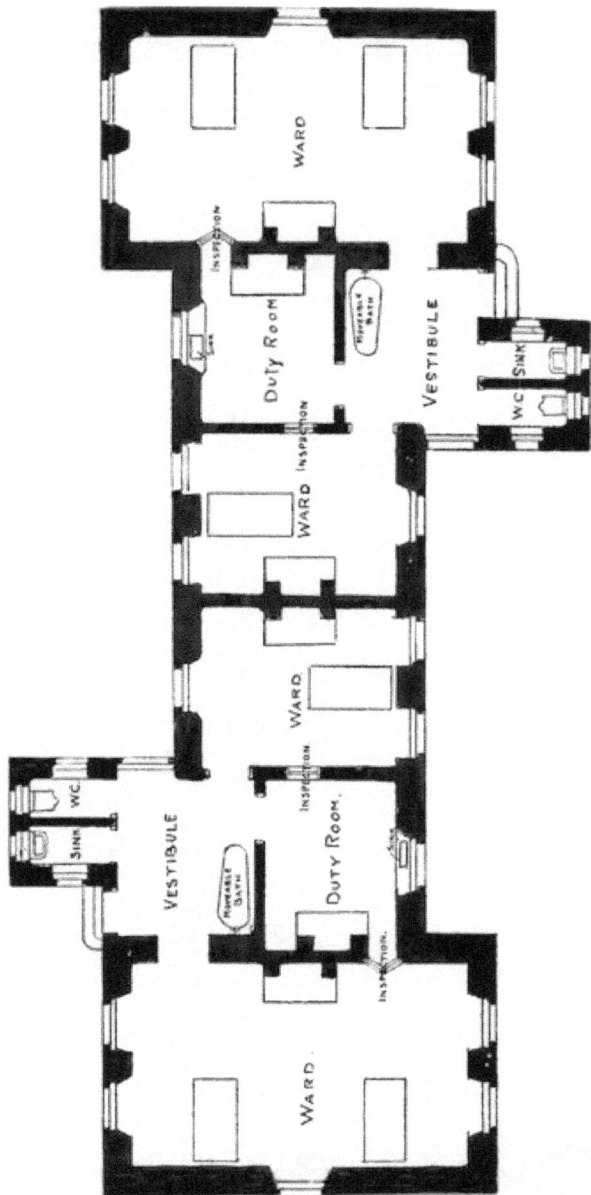

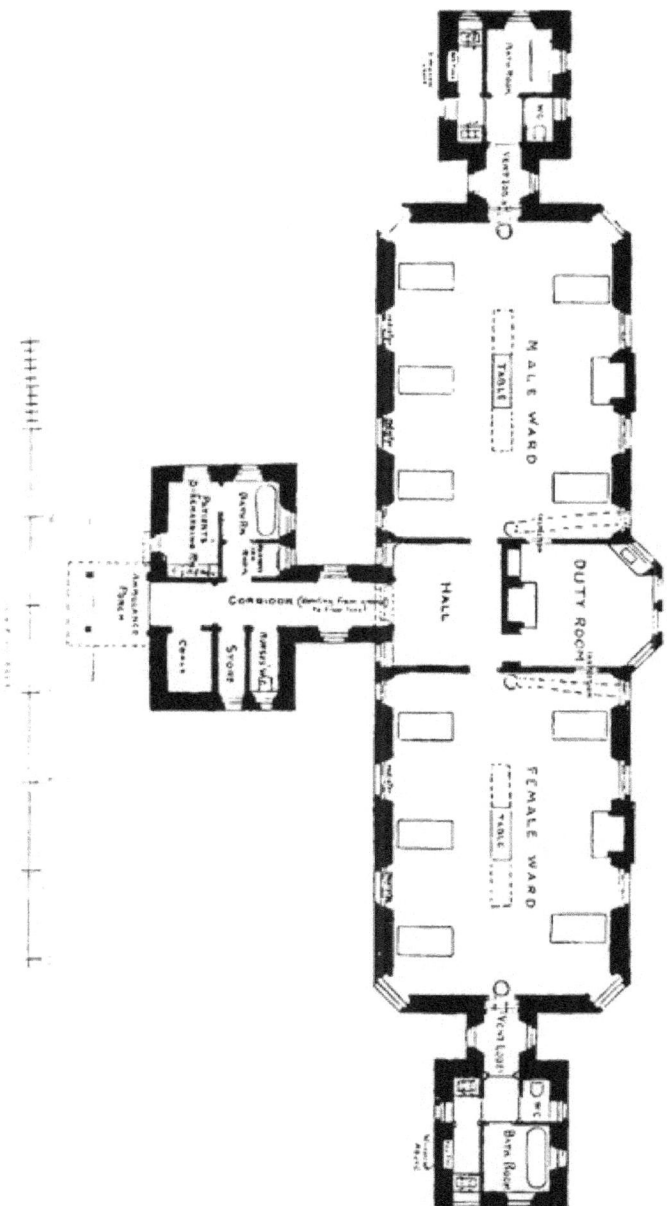

STONEHOUSE ISOLATION HOSPITAL, LANARKSHIRE.

FEVER WARDS.

MALE WARD

TABLE

DUTY ROOM

FEMALE WARD

TABLE

HALL

CORRIDOR (leading from wards to Fever Ward)

Patient's Dressing Room

Bath Rm

AMBULANCE PORCH

Store

Nurse W.C.

Cross

THE construction of the hospital for the borough of Sheffield was completed towards the end of 1880.

The site consists of a quadrilateral piece of land, measuring somewhat over an acre and a third, and occupying the summit of one of the numerous undulations which characterise Sheffield. It is situated about 400 yards from the centre of the borough, which covers an area of some 30 square miles. The soil is clay.

As may be seen from the ground plan, the hospital consists of —(1) An administrative block three stories high in front; (2) Four ward pavilions, two stories high, with flat roofs which are railed in, and utilised as airing grounds for convalescents; (3) Two porters' lodges; (4) Outhouses containing laundry with ironing-room and drying-room, a disinfecting chamber, a mortuary and post-mortem-room, an ambulance shed, and stabling.

The administrative block and wards are built of red brick, with stone ornamentation. The number of beds is 16 in each pavilion, or 64 in all.

The ward ceilings are flat, and there are nowhere any projections favouring the accumulation of dust. The inner facing of the walls is of cement, coloured with some wash which can frequently be renewed. The floors consist of pine planks, laid without any interspaces. The remainder of the woodwork throughout the building is of oak. The centre of each ward is occupied by a large stove, containing two open fireplaces. Fresh air, which is conveyed from without by means of special shafts under the floor on each side of the ward, passes round the fireplaces and, after being warmed, into the wards at a height of about 6 feet above the floor. Another opening near the ceiling carries some of the vitiated air into the smoke flue. Above each bed are fitted two Sherringham ventilators having direct communication with the outer air by means of shafts rising from below. The windows are arranged so that there is one near each corner of the ward. The lower three fourths of each window consists of double hung sashes,

and the upper quarter of a pivot hung frame, which easily opens
by means of Beanland's patent quadrant.

The population in 1881 was estimated at 285,621, and the
ratable value at £934,320.

The total expenditure incurred in the construction of the
hospital (including the preparation of the site, the prizes for designs
etc.), was as follows :—

Purchase money of site and interest thereon .	£1,808	8	2
Expense of conveyance . .	8	17	4
Cost of preparing site .	1,816	4	0

Cost of Construction.

Preliminary expenses .	110	14	9
Construction of building, etc.	16,092	15	3
Clerk of works . .	257	4	11
Architect's commission .	908	4	11
Disinfecting apparatus .	100	0	0
	£21,102	9	4

The above particulars, as well as the plans, were copied by
permission of Dr. Thorne Thorne, from the Supplement to the
Tenth Annual Report of the Local Government Board, re-issued
1893, pp. 241-243, where further information may be obtained.

HOSPITAL FOR INFECTIOUS DISEASES, SHEFFIELD.

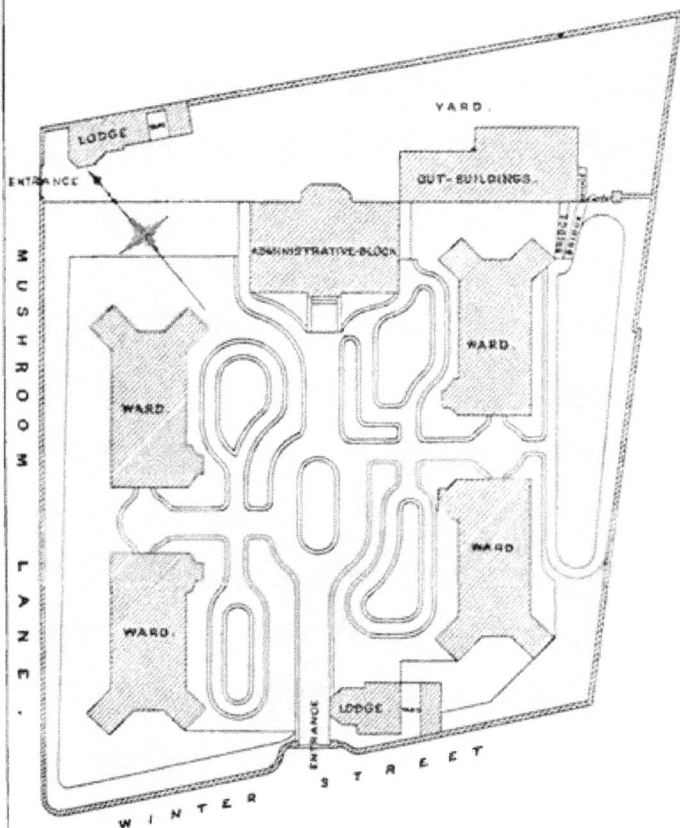

LODGE

YARD.

ENTRANCE

OUT-BUILDINGS.

MUSHROOM LANE.

ADMINISTRATIVE BLOCK

WARD.

WARD.

WARD.

WARD.

LODGE

ENTRANCE

WINTER STREET

BLOCK — PLAN.

SCALE OF FEET.

10 5 0   10 20 30 40 50   100   150   200

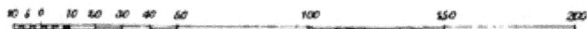

S.L. Swann Archt.
Sheffield Dec 16.80.

HOSPITAL FOR INFECTIOUS DISEASES, SHEFFIELD.

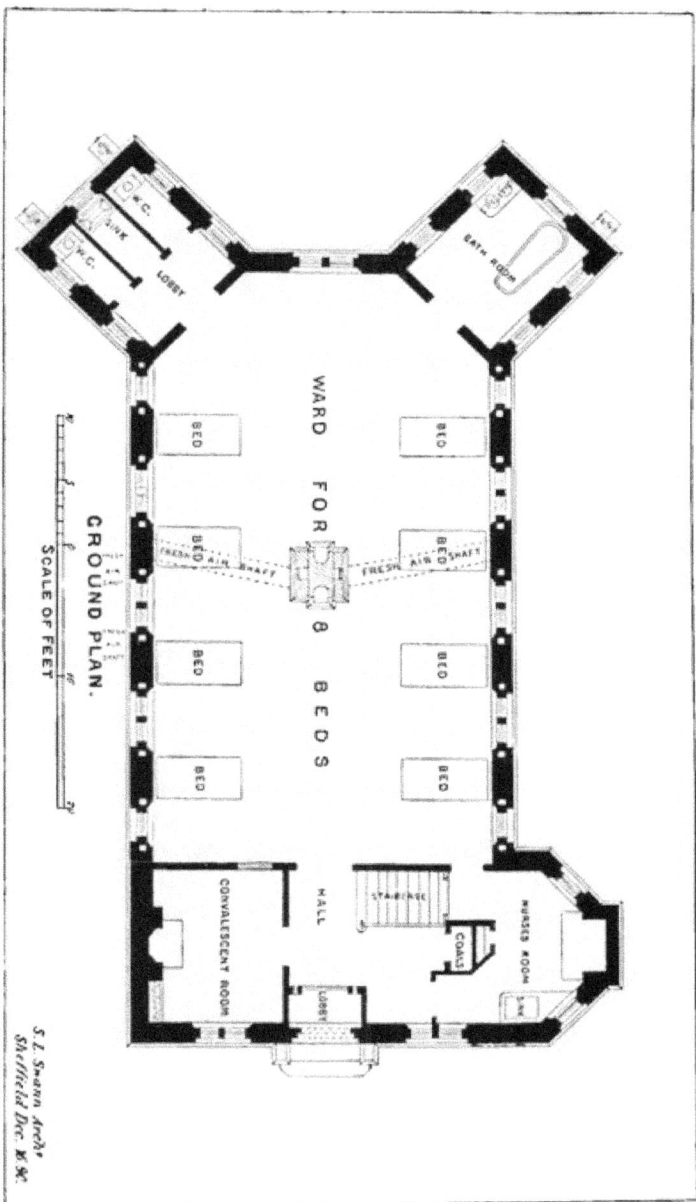

WARD FOR 8 BEDS

BED
BED
BED
BED
BED
BED
BED
BED

FRESH AIR SHAFT
FRESH AIR SHAFT

W.C.
W.C.
SINK
LOBBY

BATH ROOM

CONVALESCENT ROOM

HALL

STAIRCASE

COALS

NURSES ROOM

LOBBY

GROUND PLAN.

SCALE OF FEET

S. L. Swann Arch.
Sheffield Dec. K.K.

HOSPITAL FOR INFECTIOUS DISEASES, SHEFFIELD.

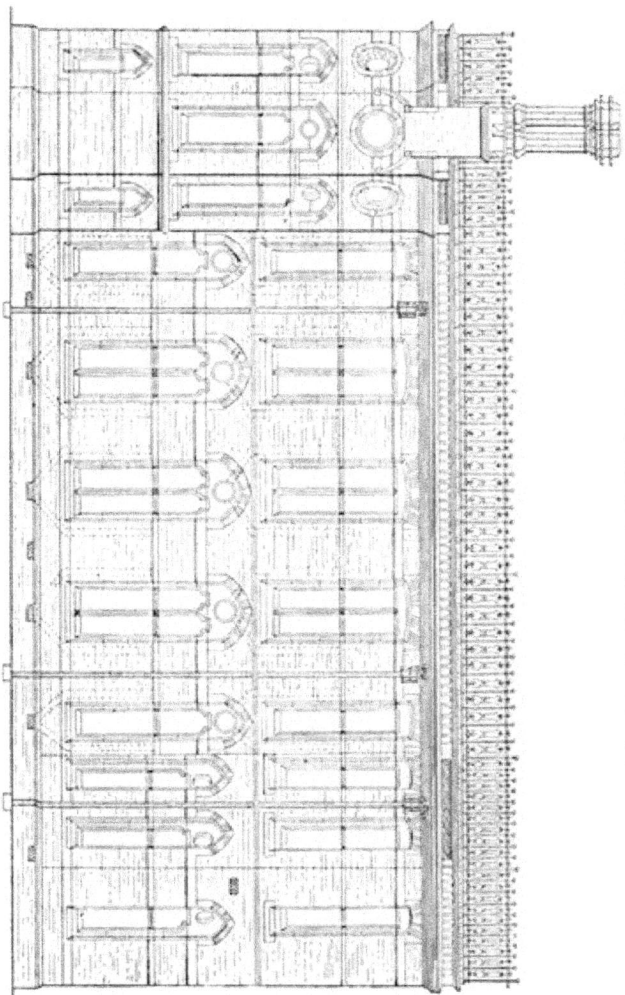

ELEVATION.
SCALE OF FEET.

S. L. Swann Arch.
Sheffield Pack Rd.

DUNOON AND KILMUN ISOLATION
HOSPITAL

THE following plans were submitted to the Dunoon Combination Hospital Committee for the Isolation Hospital intended to be erected for the burgh of Dunoon and the landward parts of the parishes of Dunoon and Kilmun. Sketch A was selected by the committee as meeting the requirements of the locality.

The site selected is about two acres in extent on a slope of porous soil overlying clay slate, with a south-eastern exposure. It is within a mile of the centre of the burgh, which covers an area of about 1100 acres, and is also in the most central position for the whole district to be served by the hospital.

The area of the whole district is about 44,577 acres. The population in 1891 was 8683, of whom 5283 reside in the burgh of Dunoon, the majority of the remaining 3400 being in villages along the coast, within 8 miles of the proposed hospital. The population is probably doubled during the summer months. A few scattered families live at distances up to about 25 miles from the hospital.

The ratable value of the district is about £82,000.

As may be seen from Sketch A, the proposed hospital is to consist of—(1) An administrative building and outhouses; (2) One detached pavilion with two wards of four beds each, intended for the reception of persons suffering from the more prevalent diseases, such as scarlet fever; (3) Two detached smaller pavilions or cottages with two wards each, each ward having sufficient room for two patients. These are intended to be used for the less prevalent diseases, such as diphtheria or typhoid fever, or as private wards, as circumstances may require.

The number of beds provided is 16, and the building is so arranged that persons of both sexes suffering from three different infectious diseases may be isolated simultaneously in the hospital. The number of beds is in excess of the ordinary requirements of such a small population. This was considered necessary owing to

DUNOON·COMBINATION·
FEVER·HOSPITAL.

SKETCH A, AS INTENDED TO BE ERECTED

·BLOCK·PLAN·

PROPOSED   NEW   ROAD.

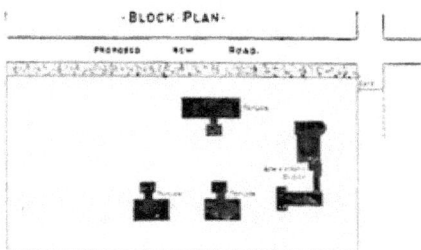

ISOLATION
WARDS

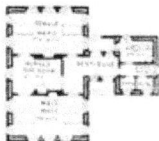

ISOLATION
WARDS

FEVER
WARDS

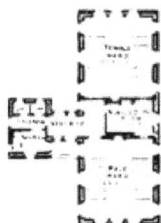

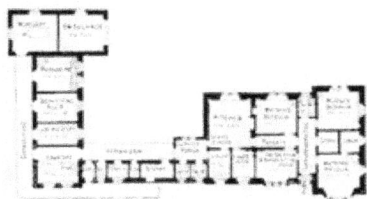

ADMINISTRATIVE BLOCK

SCALE OF FEET.

the increase of the population during the summer months. From its position in regard to a number of large towns, and being a popular seaside resort, the district is, like other watering places, liable to become frequently infected by visitors suffering or recovering from infectious diseases.

The large wards in the fever pavilion are 26 feet long by 22 feet broad. The small wards in the isolation pavilions are 22 feet long by 13 feet broad. A height of about 14 feet is therefore necessary in order to provide a cubic space of 2000 feet per patient.

The following are the dimensions of the other principal apartments :—Nurses' day-rooms 15 feet by 10 feet, kitchen 15 feet 6 inches by 13 feet, doctor's room 13 feet by 9 feet, matron's parlour 13 feet by 11 feet, 2 bedrooms 13 feet by 11 feet, 1 bedroom 9 feet by 7 feet, mortuary 14 feet by 11 feet, ambulance shed 14 feet by 11 feet, washing-house 13 feet by 11 feet, disinfecting-room 13 feet by 11 feet.

In Sketch B the pavilions are also three in number, and contain two wards with two beds in each. The future extension of the wards was intended to be carried out in a different manner.

After this was accomplished, each pavilion would consist of a male and female ward of two beds each, for acute cases, and of a male and female ward of three beds each for convalescents, raising the total number of beds to thirty.

The wards in this plan are 20 feet long by 15 feet broad. A height of about 13 feet 6 inches is therefore necessary in order to provide a cubic space of 2000 feet per patient. The dimensions of the other principal apartments are as follows, viz. :—Nurses' day-rooms 12 feet 6 inches by 9 feet, kitchen 15 feet by 12 feet, matron's parlour 13 feet by 11 feet, doctor's room 13 feet by 10 feet, 2 bedrooms 13 feet by 11 feet, 2 bedrooms 11 feet by 11 feet, 1 bedroom 9 feet 6 inches by 6 feet, mortuary 14 feet by 9 feet 6 inches, ambulance shed 14 feet by 10 feet, disinfecting-room 12 feet by 9 feet, washing-house 12 feet by 12 feet, laundry 12 feet by 12 feet.

There is also a discharging-room, and store for clean clothes, and nurse's dressing-room detached both from the wards and from the administrative building. This building is intended for the final baths and disinfection of patients before being sent to their homes. It is also intended to be used as a dressing-room for nurses before going on duty, and after leaving the wards.

In the discharging-rooms the dimensions are as follows, viz.—Undressing-rooms (*a*) 10 feet by 8 feet, (*b*) 10 feet by 6 feet, (*c*) 11 feet by 10 feet, shower bath-room 10 feet by 4 feet 6 inches, patients' bath-room 10 feet by 9 feet, nurses' bath-room 10 feet by 10 feet, patients' dressing-room, 10 feet by 8 feet, nurses' dressing-room 10 feet by 8 feet, clothes store 18 feet 6 inches by 8 feet.

The whole building is to be constructed of stone and lime, with hollow walls.

The estimated cost is as follows :—

Plan A.

1 Pavilion, 8 beds		£815	0 0
1 „ 4 „		615	0 0
1 „ „		615	0 0
Administration block and outhouses		1010	0 0
		£3055	0 0
Laying off grounds	£30 0 0		
Enclosures	120 0 0		
Furniture	175 0 0		
		325	0 0
		£3380	0 0

Plan B.

3 Pavilions, each of 4 beds, £630 each .	=	£1890	0 0
Administrative block and outhouses	.	1200	0 0
	.	£3090	0 0
Laying off grounds	£30 0 0		
Enclosures	120 0 0		
Furniture	140 0 0		
		290	0 0
Additional extension, 6 beds, as per Plan, to each pavilion, £260, or, 18 beds . .		1560	0 0
		£4940	0 0
Discharging-room	.	500	0 0
30 beds at £181 : 6 : 8 .		£5440	0 0

For the plans and estimate of cost I am indebted to Mr. Bryden, the architect.

·DUNOON· COMBINATION·
·FEVER· HOSPITAL·

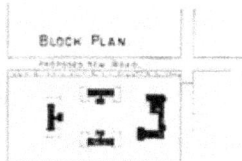

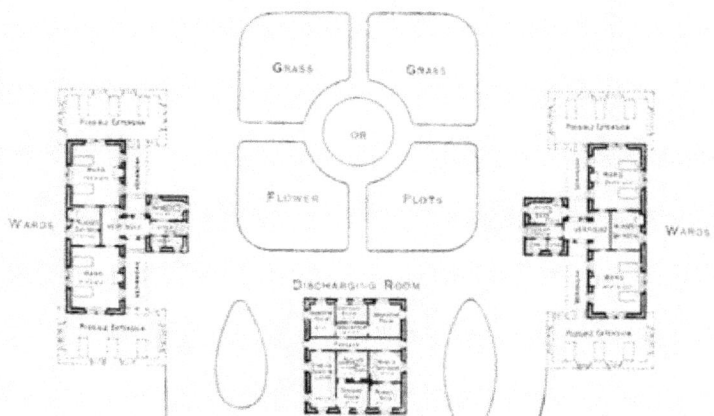

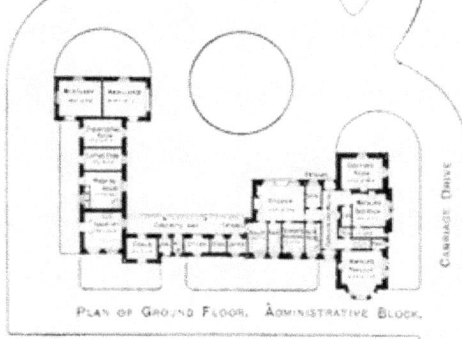

"SKETCH B".

WARDS

BLOCK PLAN

PROPOSED NEW ROAD

GRASS   GRASS

OR

FLOWER   PLOTS

WARDS

WARDS

DISCHARGING ROOM

CARRIAGE DRIVE

PLAN OF GROUND FLOOR. ADMINISTRATIVE BLOCK.

PLAN OF UPPER FLOOR

SCALE OF FEET

PLANS OF ISOLATION HOSPITALS RECOMMENDED BY THE LOCAL GOVERNMENT BOARD

On the Provision of Isolation Hospital Accommodation by Local Sanitary Authorities

ENGLISH communities nowadays recognise the advantage of Isolation Hospitals as a means of preventing the spread of infectious diseases from persons who cannot be properly isolated in their own homes. But too often the provision of such hospitals is put off until some infectious disease is immediately threatening or has actually invaded a district. It cannot be too clearly understood that an Isolation Hospital, to fulfil its proper purpose of sanitary defence, ought to be in readiness beforehand. During the progress of an epidemic it is of little avail to set about hospital construction. The mischief of allowing infection to spread from first cases will already have been done, and this mischief cannot be repaired. Thus, hospitals provided during an epidemic are mainly of advantage to particular patients; they have little effect in staying the further spread of infection. Moreover, hospitals provided under such circumstances, to be of any use, must be large and costly; and their construction can seldom be of a kind that is suited in after times for the isolation requirements of their districts.

The present memorandum is designed to represent to every Sanitary Authority which is without means of isolation for first cases of infectious sickness appearing in its district, the importance of providing itself against that event, and of doing so before the invasion of actual infection. It is intended also to suggest to Sanitary Authorities of rural districts, and of small towns, the means by which they may most advantageously make such provision. Some general principles to be held in view by all authorities who are about to establish Isolation Hospitals for their districts will be illustrated in the course of the memorandum.

As regards villages.—Large villages and groups of adjacent villages will commonly require the same sort of provision as towns. Where good roads and proper arrangements for the conveyance of the sick have been provided, the best arrangements for village populations is by a small building accessible from several villages; otherwise the requisite accommodation for (say) four cases of infectious disease in a village may at times be got in a fairly isolated and otherwise suitable four-room or six-room cottage at the disposal of the Sanitary Authority; or by arrangement made beforehand with some trustworthy cottage-holders, not having children, that they should receive and nurse, on occasion, patients requiring such accommodation.

In towns, hospital accommodation for infectious diseases is wanted more constantly, as well as in larger amount, than in villages; and in towns there is greater probability that room will be wanted at the same time for two or more infectious diseases which have to be treated separately. The permanent provision to be made in a town should consist of not less than four rooms in two separate pairs; each pair to receive the sufferers from one infectious disease—men and women of course separately. The number of cases for which permanent provision should be made must depend upon various considerations, among which the size and the growth of the town, the lodgment and habits of its population, and the traffic of the town with other places, are the most important. There is no fixed standard therefore by which the standing hospital requirements proper for a town can be measured. Furthermore, it is to be remembered that occasions will arise (as where infection is brought into several parts of the town at one time) when isolation provision, in excess of that commonly sufficient for the town, will become needful.

For a town the hospital provision ought to consist of wards in one or more permanent buildings, with space enough for the erection of other wards, temporary or permanent. Considerations of ultimate economy make it wise to have permanent buildings sufficient for somewhat more than the average necessities of the place, so that recourse to temporary extensions may less often be necessary. And in any case it is well to make the administrative offices somewhat in excess of the wants of the permanent wards; because thus, at little additional first cost, they will be ready to serve, when occasion comes, for the wants of temporary extensions.

Plans illustrating the sanitary requirements of small hospitals for infectious disease are arranged on three sheets accompanying the present memorandum. Plan A, on the first sheet, is that of a little building to hold two patients of each sex. On the second sheet a plan and a section (B)[1] of a rather larger hospital building are shown, providing for eight patients, with separation of sex, and also of one infectious disease from another. A convenient disposition of buildings upon site is also indicated on the same sheet. The third sheet shows a plan and section (C) of a small pavilion adapted to receive six male and six female patients suffering from one kind of infectious disease. It will be found that in all the plans proper standards of space are observed, viz., not less than 2000 cubic feet of air space, than 144 square feet of floor space, and 12 linear feet of wall space to each bed ; that means are provided for the adequate ventilation and warming of wards, and for securing them from closet emanations and the like. In plan A, earth-closets, in other plans water-closets, are indicated as the means of excrement disposal. The latter are to be regarded as preferable where efficient sewers are available. Places for washing and disinfection, and for a mortuary, are indicated. It will be observed that an interval of 40 feet is everywhere interposed between every building used for the reception of infected persons or things and the boundary of the hospital site. This boundary should have a close fence of not less than 6 feet 6 inches in height, and the 40 feet of interval should not afterwards be encroached on by any temporary building or other extension of the hospital. In the construction and arrangement of such temporary buildings as may at times be wanted in extension of the permanent hospital, the same principles should be held in view.

In determining the locality where an infectious hospital should be placed, the wholesomeness of the site, the character of the approaches, together with the facilities for water supply, and for slop and refuse removal, are matters of primary importance.

[Sites for hospitals designed to receive smallpox require a very much larger space about them than sites for other infectious-diseases hospitals. Smallpox hospitals, as we know them, are apt to disseminate smallpox, and their sites should therefore be placed outside of towns, and should indeed be sought at places as far distant from any populated neighbour-

---

[1] On the second sheet, B, plans and sections of hospital buildings are shown providing for 6 and 10 instead of 8 patients, as stated above. These plans also are recommended by the Local Government Board.

hood as considerations of accessibility permit. It has been suggested that smallpox hospitals may be so constructed as not to be dangerous to neighbouring habitations; and that this can be done by a system of passing through a furnace all outgoing air from infected wards and places.]

Useful information on points of construction and administration of Isolation Hospitals, derived from experience of them in various parts of England and Wales, will be found in a report (C.—3290) of the Medical Department, 1882—re-issued in 1884.

R. Thorne Thorne,
Medical Officer.

Local Government Board, Medical Department,
September 1892.

PLAN A.

ELEVATION.

N.B. Moveable baths and bath Commodes will be expressed for the wards. When nurses bedrooms are not provided in the over-looker's cottage, they may be placed in an upper storey of the ward block not shown in the Elevation.

10.1 This distance should be to too

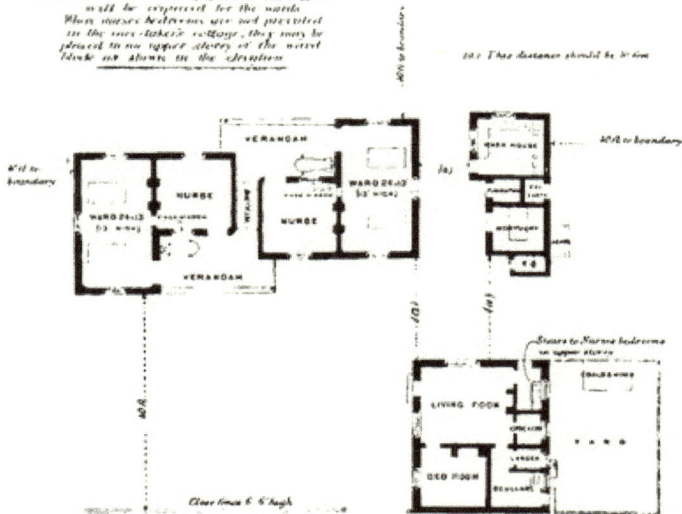

PLAN.

Scale 16 ft. to one inch.

VERANDAH
NURSE
WARD 24×13 5 beds
WARD 24×13 5 beds
NURSE
VERANDAH

Well to boundary

40 ft.

Clear fence 6' 6" high

WASH HOUSE
40 ft. to boundary

Stairs to Nurses bedrooms in upper storey

LIVING ROOM
COAL & WOOD
BED ROOM
YARD

PLAN B.

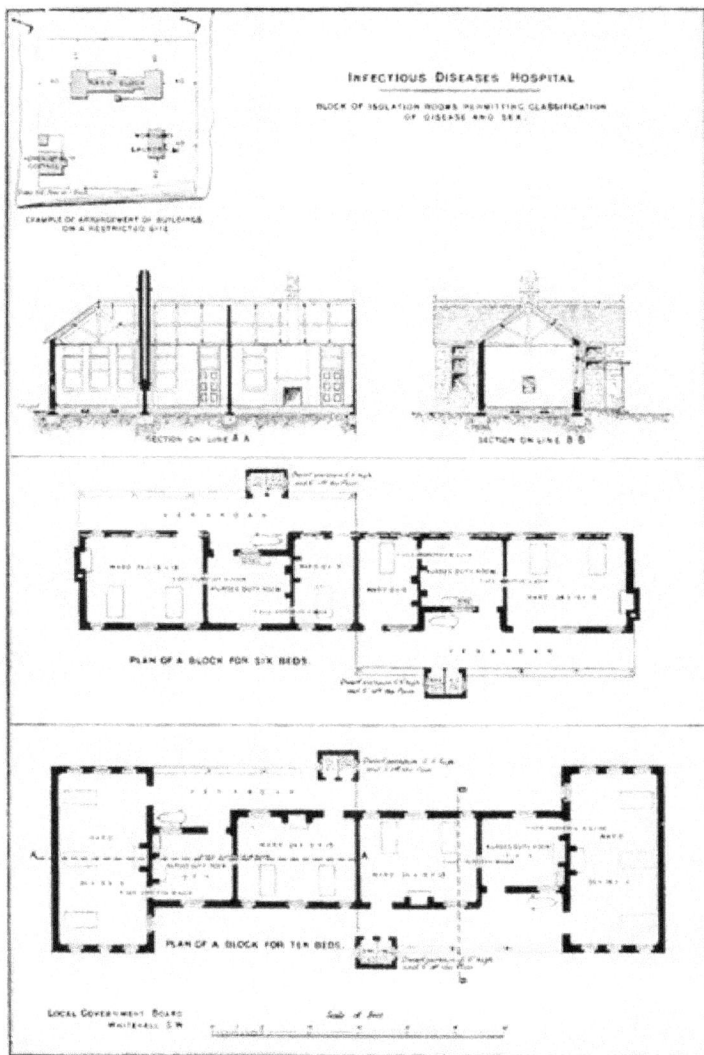

INFECTIOUS DISEASES HOSPITAL

BLOCK OF ISOLATION ROOMS PERMITTING CLASSIFICATION
OF DISEASE AND SEX.

EXAMPLE OF ARRANGEMENT OF BUILDINGS
ON A RESTRICTED SITE.

SECTION ON LINE A A

SECTION ON LINE B B

PLAN OF A BLOCK FOR SIX BEDS.

PLAN OF A BLOCK FOR TEN BEDS.

LOCAL GOVERNMENT BOARD
WHITEHALL S.W.

Scale of feet

PLAN C.

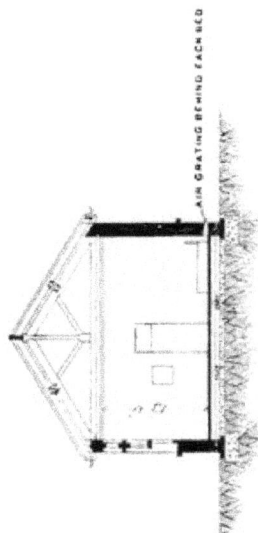

SECTION ON LINE A.A.

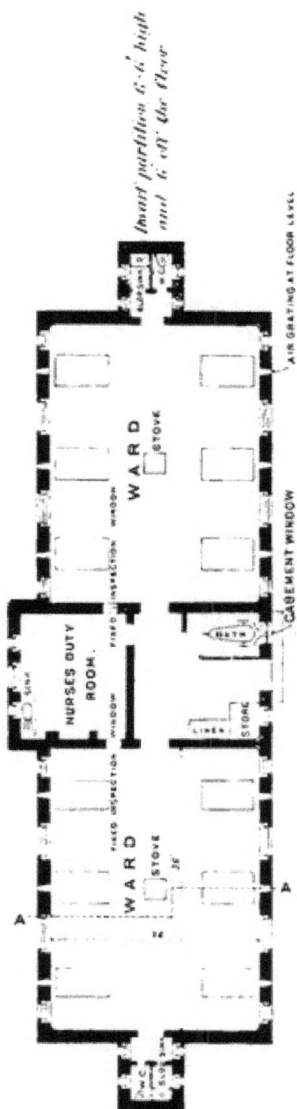

AIR GRATING BEHIND EACH BED

WARD

STOVE

Short partition 6·6 high
and 6 off the floor

SLOP SINK

W.C.

AIR GRATING AT FLOOR LEVEL

CASEMENT WINDOW

BATH

LINEN

STOVE

NURSES DUTY
ROOM.

SINK

FIXED INSPECTION WINDOW

FIXED INSPECTION WINDOW

WARD

STOVE

A

A

PLAN OF A WARD PAVILION FOR 12 BEDS.

TARBERT ISOLATION HOME

THE following plan was submitted for a hospital which is intended to be erected at Tarbert, Loch Fyne. It consists of an administrative cottage, wards, and outhouses, all detached, and at safe distances apart. There is also space left for additional extension, should that be found necessary in the future. The ward block consists of four rooms. If only one disease be under treatment in the hospital, three rooms may be occupied by patients, and the remaining room by the nurse in attendance. If two diseases such as typhoid fever and diphtheria break out at the same time in the district, two patients suffering from one disease might be isolated in one ward, and two patients from the other disease in the other ward. By closing the door of the porch in the verandah, all communication between the two ends of the hospital would be prevented, and a room left for the nurses attending to each disease.

The site selected for the hospital is within half a mile of the village of Tarbert. The area of the whole district to be served by the hospital is about 50,000 acres. The population is about 3000, of whom 1900 reside in the village of Tarbert. The remainder are scattered over the district at distances by road up to about fifteen miles.

The ratable value of the district is about £14,745.

The estimated cost of the building is, if of stone and lime, £1050. If the wards were built of stone, but the administrative cottage and outhouses of cement, the cost is estimated at £850.

For the plans and estimate of cost I am indebted to Mr. Petrie, the architect.

TARBERT ISOLATION HOME.

DRYING GREEN.

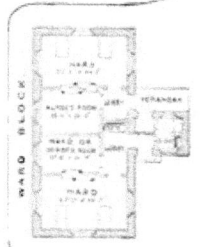

WARD BLOCK

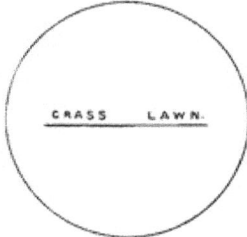

PLAN OF OFFICES.

GRASS LAWN.

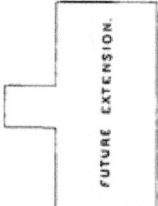

FUTURE EXTENSION.

ADMINISTRATIVE BLOCK.

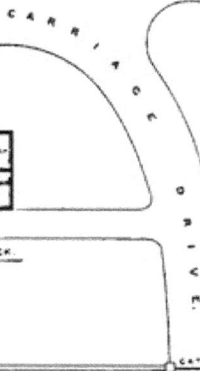

CARRIAGE DRIVE.

GATE

PLAN OF GROUND.

SCALE OF FEET.

Alex Petric
Architect.
131 Willington St
Glasgow.

DR. BURDON SANDERSON'S ANNULAR
WARD FOR SMALLPOX

DR. BURDON SANDERSON'S ANNULAR WARD FOR SMALLPOX

Annular Ward for the reception of 12 Smallpox patients

THE following are architect's sketches of annular ward referred to in Dr. Burdon Sanderson's evidence before the Smallpox and Fever Hospitals Commission. "The entrance from the corridor is closed by double spring doors. The corridor communicates (1) with the exterior, (2) with the mortuary and disinfecting chamber, (3) with chambers for changing the clothes of visitors and nurses ; consequently the corridor is regarded as infected. If desirable it might be brought within range of the ward ventilation, and used as a place of recreation for patients sufficiently advanced in recovery. In a building consisting of several stories, the ward might be entered from a staircase in the central chamber, the construction of which would not interfere with the ventilation."—Smallpox and Fever Hospitals Commission Report, Minutes of Evidence, Questions 5488, 5492.

ANNULAR WARD FOR THE RECEPTION OF 12 SMALLPOX PATIENTS.

DR. J. BURDON SANDERSON, F.R.S.*

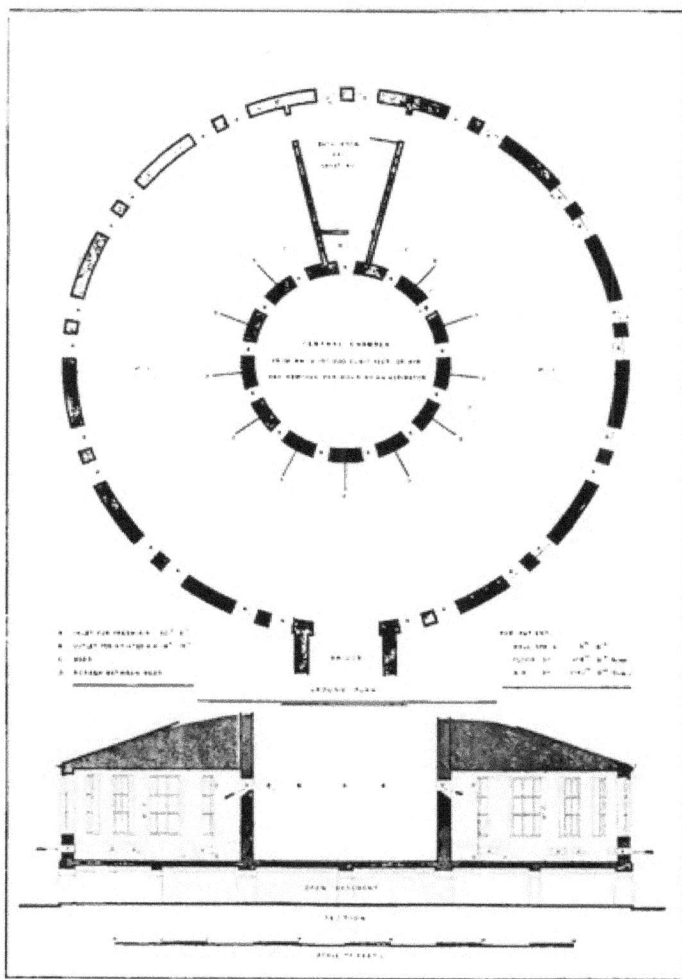

* Smallpox and Fever Hospitals Commission, Minutes of Evidence, Question 5482.

INDEX

THE END

Printed by R. & R. CLARK, *Edinburgh*

HOSPITALS

COVERED INSIDE AND OUTSIDE WITH

WIRE-WOVE WATERPROOF ROOFING

Are Proof against Heat and Cold, give
no harbour for Infection, and can be
washed down with a Hose from Roof
to Floor.

**SAMPLES, DESCRIPTIVE PAPERS, AND ESTIMATES
FOR ALL KINDS OF BUILDINGS FROM**

THE PATENT WIRE-WOVE
WATERPROOF ROOFING COMPANY,
LIMITED,

108 Queen Victoria Street, E.C.

The "GREENALL" Steam Washer.

www.ingramcontent.com/pod-product-compliance
Lightning Source LLC
Chambersburg PA
CBHW021503210326
41599CB00012B/1120